The Chemistry of Fermentation Processes: From Food to Fuel

THE CHEMISTRY OF FERMENTATION PROCESSES: FROM FOOD TO FUEL

Edited by

Dr. Tirth

ARCLER PRESS

www.arclerpress.com

The Chemistry of Fermentation Processes: From Food to Fuel

Dr. Tirth

Arcler Press
4164 Lakeshore Road
Burlington, ON L7L 1A4
Canada
www.arclerpress.com
Email: orders@arclereducation.com

ISBN: 978-1-77956-414-6 (Hardcover)

ABOUT THE EDITOR

Dr. Tirth is presently working as an assistant professor in the department of Chemistry, Faculty of Sciences, Parul University. He did his M.Sc. & Ph.D. in Chemistry from the Maharaja Sayajirao University, Vadodara. His research interest is in Drug Discovery, Synthetic Organic Chemistry and Natural Products isolation. To add to her credits, he has published 15 articles in various international journals. He is having 4 U.S. & 1 Indian patents so far.

TABLE OF CONTENTS

LIST OF FIGURES

LIST OF TABLES

PREFACE

Fermentation has been a cornerstone of human advancement, influencing the food we consume, the beverages we enjoy, and the renewable fuels that are increasingly vital in today's world. Fermentation processes are among the oldest culinary technologies, created to improve food storage stability and texture. Fermented foods continue to be popular today, accounting for an important amount of daily consumption. Important examples include bread-making; dairy products such as yogurts, cheeses, buttermilk, and sour milk; alcoholic drinks, such as wine, beer, and cider; fermented vegetables, such as sauerkraut and pickles; and fermented meats, such as sausages and salami.

The book's discussion starts with the introduction of fermentation, outlining its definition, biological roles, and biochemical foundations. It delves into the principles governing fermentation and categorizes, emphasizing the pivotal role of microorganisms and enzymes. Technological advancements that optimize fermentation processes are discussed, highlighting their significant impact on food and fuel applications. Future challenges and potential developments in the field will also be considered. Chapter 2 examines primary and secondary metabolites as biotechnological products, detailing the fermentation processes and the role of fermentative microorganisms. Fermented foods' benefits and potential side effects are explored, providing a balanced view of their impact on human health.

Chapter 3 focuses on the specific health benefits associated with fermented meat and vegetable products, indigenous fermented foods, and functional foods, including probiotics, underscoring their health benefits and mechanisms of action. Chapter 4 addresses bioethanol production from molasses biomass, focusing on the different types of molasses and the general process for ethanol production. Specific fermentation techniques using various yeast strains and microorganisms are investigated, along with strategies to enhance ethanol production efficiency.

Chapter 5 analyzes the role of fermentation in altering the color of foods, providing insights into the biochemical reactions responsible for these changes in meat, fish, dairy, and vegetables. Chapter 6 covers gas fermentation for biofuel production, detailing the critical parameters that influences the process, such as gas composition, substrate pressure, medium formulation, pH, and temperature. The types of gasifiers used, and the biochemistry and microbial aspects of gas fermentation are also discussed. Chapter 7 shifts the focus to fermented cereal and legume products, describing their types, manufacturing processes, and the microorganisms involved. The chemical changes induced by pre-fermentation and fermentation processes are explained, providing a comprehensive understanding of the biochemical transformations that occurs.

Chapter 8 includes the practical applications of fermentation, highlighting the benefits and challenges of lactic acid fermentation. The importance of fermentation in public health is emphasized, along with practical interventions to enhance the safety of fermented foods, illustrated by a HACCP study of Togwa.

This book has been written specifically for students and scholars to meet their needs in terms of knowledge and to provide them with a broad understanding of business information systems.

—Author

CHAPTER 1

INTRODUCTION TO FERMENTATION

LEARNING OBJECTIVES

At the end of this chapter, readers will be able to:

1. Define fermentation and its historical significance in food production.
2. Classify biochemical processes involved in fermentation.
3. Explain various types of fermentation and their applications.
4. Understand the role of microorganisms and enzymes in fermentation.
5. Explore principles and mechanisms of fermentation.
6. Value the impact of technology on optimizing fermentation processes.
7. Know the challenges and future directions of fermentation in various industries.

INTRODUCTORY EXAMPLE

In the realm of the production of food, fermentation is a fundamental technique that plays a crucial role in converting raw ingredients into delicious and healthy products. Consider, for example, the well-known artisanal cheese company, "Rustic Creamery." This company is dedicated to creating high-quality artisanal cheeses using time-honored fermentation techniques that have been handed down from one generation to the next.

Rustic Creamery selects premium milk of nearby farms and utilizes a precise fermentation techniques to craft its unique assortment of cheeses. By harnessing the precise workings of lactic acid-producing bacteria like *Lactobacillus*, the milk goes through a process of fermentation, for the creation of curds and whey. The curds undergo a pressing process, followed by a meticulous aging and monitoring phase to cultivate the distinct flavors and textures that define each type of cheese.

Through a combination of traditional methods and cutting-edge techniques, Rustic Creamery has established itself as a renowned producer of high-quality cheeses that are highly regarded by both experts and passionate cheese lovers. Their steadfast dedication to excellence and environmental responsibility highlights the lasting importance of fermentation in modern culinary practices.

UNIT INTRODUCTION

Through a combination of traditional methods and cutting-edge techniques, Rustic Creamery has established itself as a renowned producer of high-quality cheeses that are highly regarded by both experts and passionate cheese lovers. Their steadfast dedication to excellence and environmental responsibilities highlights the lasting importance of fermentation in modern culinary practices fermentation, as currently defined, is the metabolic process in which a chemical is and fragmented down into simpler forms, typically due to the activities of microorganisms like yeast or bacteria. In the disciplineof physiology and biochemistry, fermentation processes often involve metabolic pathways that induce alterations in organic molecules through the activities of microbes and enzymes (Valyasevi & Rolle, 2002).

From a biochemical perspective, fermentation occurs when anaerobic organisms interact with organic substrates. Nevertheless, modern usage broadens the definition to include the microbiological production of smaller organic molecules, regardless of whether it occurs in the presence or absence of oxygen. In the field of applied sciences, such as food or beverage production, fermentation encompasses various processes driven by microbial activities that result in beneficial transformations to food or drink products. The products of fermentation can be extracted using the feedstock and sold as pure substances, which is a departure from the way fermentation was done in the prior work. (for example, producing ethanol instead of wine) (Bayrock & Ingledew, 2001).

Microorganisms have been recognized as the main agents responsible for fermentation. However, it is important to note that fermentation is a natural process that occurs throughout various ecosystems. From a scientific perspective, fermentation is a fundamental process that allows organisms to derive energy from molecules. It is a widely observed metabolic phenomenon across various forms of life: Microorganisms such as bacteria, archaea, and eukaryotes. Therefore, fermentation generates ATP through the breakdown of organic nutrients, typically in the absence of oxygen. Through the gastrointestinal tracts of animals, fermentation takes place due to the presence of microbial flora (Kennedy & Krouse, 1999).

Throughout history, humans have been captivated by the possibilities offered by fermentation-based processes. Tracing the beginnings of fermentation proves to be quite a challenge, as it is believed that the initial fermentation process came upon by chance when salt was unintentionally mixed with food. Fermentation techniques were traced since 7000 BC. Throughout history, people have been producing a variety of fermented foods, such/ as beer, wine, leavened bread, and cheeses. These traditional practices date back to ancient times in the Western regions. Following the decline of these cultures, East Asian region (6000–1500 BC) swiftly emerged as the next dominant civilizations, characterized by the production of a diverse range of fermented foods and beverages utilizing local techniques. This included yogurt and various other fermented dairy goods, vinegar (soured wine), pickles, butter, sauerkraut, and a variety of traditional alcoholic beverages. The objectives of utilizing fermentation-based procedures have changed over time, influenced by cultural and social factors as well as advancements in engineering and technological capabilities (Jørgensen et al., 2004).

As a result, fermentation is widely utilized across different cultures globally to enhance the preservation of food for shorter periods or to create beverages and local dishes that feature fermented sauces, vegetables, fish, or meat, thereby expanding the culinary options. This type of fermentation is termed as 'classical fermentation' or 'Indigenous fermentation.' It is utilized to create various beverages and foods by employing natural microbes or adhering to traditional procedures rooted in cultural practices (Figure 1.1). Indigenous fermentation has led to the standardization and commercialization of various products such as ales with natural yeast, cheeses with natural fungi, and wines with natural yeast. Additionally, some products are produced and sold in small quantities for specialized markets, while others remain uncommercialized, representing the unique products of indigenous and local cultures. These products, such as kefir, kimchi, and sauerkraut, are prime examples (Neubauer et al., 1991).

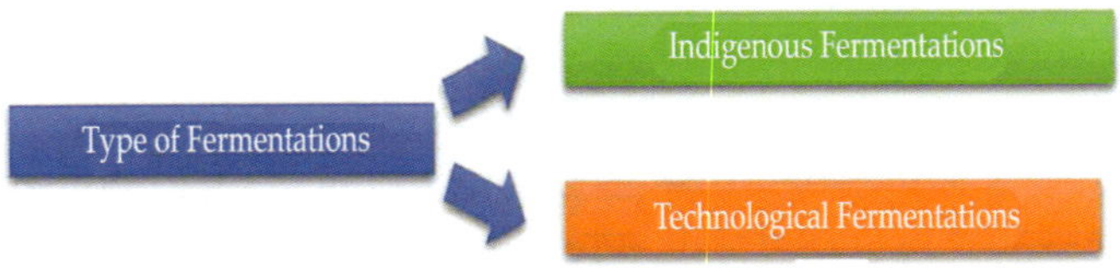

Figure 1.1. *Illustration of the classification of fermentation based on the nature of the process.*

Source: Rosa María, Creative Commons License.

1.1. OVERVIEW OF FERMENTATION

Learning Objectives:

- Understand the concept and process of fermentation.
- Understand the biological and biochemical role of fermentation.

Fermentation is a process of metabolism that brings about chemical transformations in organic substances using enzymes. In the field of biochemistry, it is specifically referred to as the process of obtaining energy from carbohydrates without the presence of oxygen. However, in the context of food production, fermentation can encompass a wider range of processes where microorganisms play a role in bringing about beneficial changes to food or beverages. Zymology is the scientific study of fermentation (Gupta et al., 2002).

Microorganisms primarily produce adenosine triphosphate (ATP) through the breakdown of organic nutrients in the absence of oxygen, known as fermentation. Fermentation has been a vital process in the production of several food and beverage items, with its origins tracing back to the Neolithic era. One use of fermentation is the preservation of food by the production of lactic acid, which imparts a sour taste to foods like yogurt, pickled cucumbers, kimchi, and kombucha. Fermentation is used in the production of alcoholic beverages such as beer and wine. Fermentation takes place in the gastrointestinal tracts of all animals, including humans (Krishna, 2005). Industrial fermentation is a more encompassing term that refers to the utilization of microorganisms in the large-scale manufacturing of various substances, such as chemicals, proteins, biofuels, enzymes, and pharmaceuticals (John et al., 2007).

1.1.1. Definitions of Fermentation

For numerous individuals, fermentation is associated with the creation of alcoholic beverages. Grains and fruits undergo fermentation to yield wine and beer. If a food becomes sour, it is often described as being spoiled or undergoing a natural fermentation process. Below are several definitions of fermentation (Dimidi et al., 2019). The range of descriptions varies from casual and everyday language to more technical and precise definitions.

- Methods of preserving food using microorganisms (general application).
- Producing alcoholic beverages or acidic dairy items involves a specific set of processes (general use).
- A large-scale microbial procedure that occurs with or without air, as commonly defined in industry.
- Anaerobic respiration is a metabolic activity that only occurs without oxygen and releases energy (taking a more scientific approach).
- A process of metabolism that releases energy from sugar or other naturally occurring molecules, without the need for oxygen or just a system for transporting electrons, and utilizes an organic molecule to be the final acceptor of electrons (very scientific).

1.1.2. Biological Role

Fermentation is an additional method for extracting energy from molecules besides aerobic respiration. This is a technique that all bacteria and eukaryotes use. It is widely regarded as the most ancient metabolic pathway, well-suited for early environments predating the existence of plant life on the planet, specifically before oxygen within the atmosphere. Nick Lane raises concerns about this proposal, highlighting that the energy released through fermentation is relatively low. This limitation prevents it from generating the necessary thermodynamic driving force for prebiotic chemistry (Mayr, 2016). The enzymes associated with fermentations that are encoded by genes may not have been present during prebiotic chemistry. Furthermore, it is worth mentioning that the distinctions observed in the process of fermentation of archaea and bacteria suggest that fermentation may have emerged at a later stage, evolving autonomously in these two forms of early life. Yeast, a type of fungus, can be found in various environments that can sustain microorganisms, ranging from fruit skins to the digestive systems of insects and mammals, even to the depths of the ocean. Yeasts undergo a process in which the breakdown of sugar-rich molecules occurs, resulting in the production of ethanol and carbon dioxide.

The fundamental processes of the fermentation are found in every cell of higher life forms. During time of intense exercise when oxygen supply gets limited, mammalian muscle undergoes fermentation, leading to the production of lactic acid. Succinate and alanine are also produced through fermentation in the invertebrates (Snyder & Bredt, 1992).

Fermentative bacteria are crucial in the production of methane , found in various environments such as sewage digesters, cattle rumens, and freshwater sediments. And generate acetate, carbon dioxide, hydrogen, formates, and carboxylic acids. Microbial consortia collaborate to convert carbon dioxide and acetate into methane. Acetogenic bacteria undergo a process where they oxidize acids, resulting in the production of additional acetate along with one of hydrogen or formates. In the realm of Archea, methanogens play a crucial role by transforming acetate into methane (Florack & Stiekema, 1994).

1.1.3. Biochemical Overview

The process of fermentation involves the reaction between NADH and an organic electron acceptor that is naturally present. Typically, pyruvate is produced from sugar via glycolysis. The reaction results in the production of oxidized NAD+ and an organic product. Common examples of this product include hydrogen gas (H_2), ethanol, lactic acid, and sometimes carbon dioxide. Nevertheless, fermentation has the potential to yield compounds that are considered more unconventional, including butyric acid and acetone. Fermentation products are typically classified as waste products, as unable to undergo further metabolism in the absence of oxygen (Christenson, 1997).

Fermentation typically takes place in an environment without oxygen. Respiration utilizes O_2, NADH, and pyruvate to produce adenosine triphosphate (ATP). This process is known as oxidative phosphorylation. This results in a significantly higher production of ATP compared to glycolysis alone. Due to certain circumstances, fermentation is not commonly utilized in the presence of oxygen (Deville et al., 2003). Nevertheless, even when plenty of oxygen available, certain strains of yeast, like *Saccharomyces* cerevisiae, have the preference for fermentation over aerobic respiration, as long as there is a sufficient amount of sugars present. This preference is commonly referred to as the Crabtree effect. Additionally, there are fermentation processes that include obligate anaerobes, which is unable to tolerate oxygen (Arunkumar et al., 2020). While yeast is responsible for the fermentation process when making ethanol in beverages such as beers and wines, it is worth noting that other agents can also be involved. Bacteria are responsible for the fermentation process involved inmaking of xanthan gum (Chambers, 1985).

1.2. PRINCIPAL AND TYPES OF FERMENTATION

Learning Objectives:

- Understand the principle of fermentation.
- Identify and describe different types of fermentation processes and their products.

1.2.1. Principal of Fermentation

The basic idea of fermentation is use of carbohydrates as an oxygen-free energy source. The process begins with the partial oxidation of glucose to pyruvate through glycolysis. Pyruvate undergoes conversion to alcohol or acid, resulting in the regeneration of NAD+. This NAD+ can then participate in glycolysis, leading to the creation of additional ATP. The energy obtained through fermentation is significantly lower compare to that obtained through aerobic respiration (Montanari et al., 1996) (Figure 1.2).

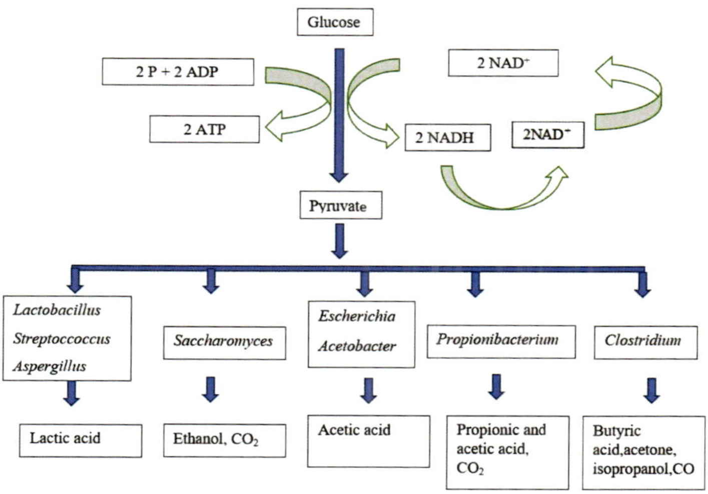

Figure 1.2. *Illustration of general pathways by which different organisms convert glucose to some fermentation end products.*

Source: Martínez-Espinosa, Creative Commons License.

- Fermentation is a biochemical process that occurs in absence of oxygen and is utilized to generate energy through the partial oxidization of glucose and other carbon sources.
- The substrate undergoes oxidation through either the Entner-Doudoroff (ED) or Embden-Meyerhoff (EMP) pathway, leading to the formation of pyruvate, ATP, and NAD (P) H.

- When external electron acceptors are unavailable, pyruvate is reduced and NAD+ is regenerated (P).
- This step is crucial for the progression of the fermentation process, resulting in the creation of ethanol and organic acids.
- ATP, the primary outcome of fermentation, is produced through substrate-level phosphorylation.
- In the second fermentation phase, NADH undergoes re-oxidation, converting back to NAD+. This process helps in reduction of pyruvate to form fermentation products like ethanol and lactate (Choi et al., 2007).
- As an illustration, in the process of glucose fermentation by streptococci lactis, the pyruvate undergoes conversion.

1.2.2. Types of Fermentation

Different types of fermentation are based on the particular microbe responsible for the reaction and the resulting product. Various types of fermentation exist:

1.2.2.1. Lactic Acid Fermentation

During this metabolic procedure, yeast strains and bacteria convert starch or sugar to a metabolite called lactate, which is lactic acid. The anaerobic process of fermentation occurs in certain bacteria and animal cells once the energy expenditure exceeds the available oxygen. During this process, a glucose molecule undergoes conversion into pyruvate and subsequently forms lactic acid (Reddy et al., 2008) (Figure 1.3).

$$CH_3CO.COO + H^+ + NADH_2 \longrightarrow CH_3CHOHCOOH + NAD^+$$

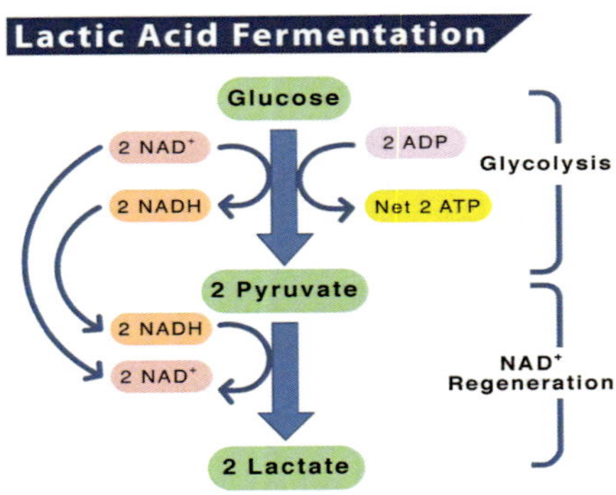

Figure 1.3. *Illustration of lactic acid fermentation.*

Source: Philip B, Creative Commons License.

1.2.2.2. Alcohol Fermentation

Ethanol fermentation, also referred as alcohol fermentation, is the enzymatic decomposition of pyruvate, a byproduct of glycolysis, into ethanol (alcohol) and carbon dioxide. Yeast performs this activity in the absence of oxygen. Alcoholic fermentation has become a process that occurs in the absence of air. Fermentation is commonly employed in the creation of alcoholic beverages such as beer, and also for the making of ethanol fuel (Aiba et al., 1968) (Figure 1.4).

$$CH_3COCOO + H^+ \longrightarrow CH_3CHO + CO_2$$

$$CH_3CHO + NADH \longrightarrow CH_3CH_2OH + NAD^+$$

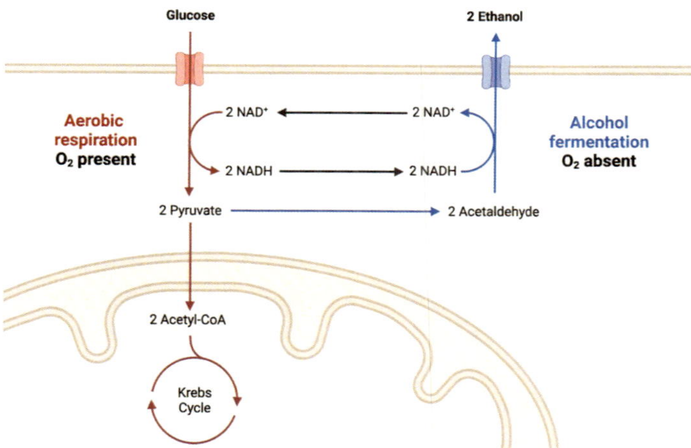

Figure 1.4. *Illustration of alcohol fermentation.*

Source: Yaakov Anker, Creative Commons License.

1.2.2.3. Acetic Acid Fermentation

Vinegar is the common name for acetic acid. The making of acetic acid typically encompasses two distinct processes. The initial step entails utilizing yeast alcohol derived from sugar, while the subsequent step involves harnessing acetic acid producing bacteria to convert ethyl alcohol into acetic acid via acetaldehyde. There are two methods used for the fermentation of acetic acid: surface fermentation and submerged fermentation (Han et al., 1992).

$$CH_3CH_2OH \longrightarrow CH_3CHO \longrightarrow CH_3C(HO)_2 \longrightarrow CH_3COOH$$

1.2.2.4. Butyric Acid Fermentation

Microorganisms break down organic compounds, such as carbohydrates, and generate butyric acid as a byproduct through their metabolic process. Butyric acid is a short-chain fatty acid that has various uses in digestion, industrial settings, and gut health (Luo et al., 2018).

$$C_6H_{12}O_6 \longrightarrow 2C_4H_8O_2 + 2CO_2 + 2H_2$$

1.3. MICROORGANISMS AND ENZYMES AS KEY TOOLS FOR FERMENTATION

Learning Objectives:

- Understand the pivotal role of microorganisms and enzymes in fermentation processes.

The nineteenth century was a time when professionals from various fields of knowledge were captivated by the idea of fermentation and its applications. The breakthrough in understanding yeast metabolic activity and its connection to fermentation can be attributed to the pioneering work of Louis Pasteur, a French chemist in the mid-nineteenth century. From that time until the start of the 20th century, a considerable amount of research has been conducted, providing detailed explanations of fermentation processes primarily influenced by microorganisms. Therefore, microorganisms are capable of carrying out anaerobic metabolic pathways, whether eukaryotic or prokaryotic, have been widely regarded as suitable candidates for fermentation-based processes (Al-Maqtari et al., 2019). Over the past 40 years, the majority of research in the field of fermentation has primarily revolved around food and beverage production, documenting natural fermentation processes, large-scale production of commercial compounds, and advancements in industrial fermentation technologies. It is important to emphasize the following aspects that could have significant implications in the fields of medicine and industries:

- The significance of microflora in mammals, such as humans and ruminants.
- The production of highly sought-after secondary metabolites by fermentation, such as pigments, antibiotics, prebiotics, and more.
- Manufacturing of biomolecules such as enzymes, amino acids, fatty acids, proteins vitamins, or surface-active compounds (Behera & Ray, 2016).

The utilization of microbes in fermentation processes at an industrial scale gained significant attention towards the end of the previous century. Several studies supported the widespread use of 'conventional' microbes in large-scale or medium-scale exploitation. During this time, several studies caught the attention of scientific community, as foccused on discovering extremophilic microbes with unique metabolic capabilities, even in oxygen-deprived environments. These microbes were seen as potential candidates for various industries, such as green biotechnology, food, and biomedicine. Hyperthermophilic microorganisms, such as members of the Archaea domain, and nonconventional yeast including *Yarrowia lipolytica* have been utilized in academic research. These microorganisms have shown potential in the production of oils and fatsIn recent years, various disciplines have made significant advancements in understanding of microbial fermentation. Therefore, two novel fields of study have arisen:

- Manufacturing of biodiesel, biogas, biofuel, and their derivatives involves a complex process that requires in-depth knowledge and expertise.
- Exploring the development and design of bioremediation and biosorption processes for soils and wastewater treatments, specifically focusing on the involvement of microbes or their isolated enzymes (Pan et al., 2022).

Microorganisms and their enzymes have played a crucial role in advancing fermentation-based processes. Numerous processes focus in the generation of microbial enzymes through fermentation, which are then utilized on a large scale, primarily in the field of biotechnology. For instance, the utilization of lignocellulose as a plentiful and environmentally friendly carbon source for the production of bioproducts has been made possible through the utilization of hemicellulolytic and cellulolytic enzymes generated by various microorganisms, such as bacteria, filamentous fungi, and yeasts. Enzymes derived from extremophilic microorganisms have shown great potential for use in the various industrial processes. One of the reasons for considering the potential use of these enzymes is their ability to maintain high catalytic efficacy even under extreme conditions. These conditions may include high temperatures, high ionic strength, extreme pH levels or limited water availability (Gurung et al., 2013).

Practice Problem

Pacific Pharmaceuticals aims to develop a more cost-effective and sustainable method for producing a key pharmaceutical compound through fermentation. As a process engineer at Pacific Pharmaceuticals, you are tasked with optimizing the fermentation process to increase yield and reduce production costs. How would you approach this challenge?

Solutions to Practice Problem

To optimize fermentation at Pacific Pharmaceuticals, analyze key parameters like temperature and pH for optimal microbial growth and product formation. Utilize advanced bioreactor technology with automated controls to regulate fermentation conditions in real-time, minimizing resource consumption while maximizing yield. Additionally, apply genetic engineering to enhance microbial strains' productivity and metabolic efficiency by boosting key enzyme expression. These strategies ensure Pacific develops a cost-effective, sustainable fermentation process for pharmaceutical production, ensuring competitiveness in the industry.

1.4. OPTIMIZING FERMENTATION PROCESSES: IMPACT OF TECHNOLOGY ON FERMENTATION

Learning Objectives:

- Understand the impact of technology on optimizing fermentation processes.

The integration of technology with fermentation has enabled the efficient and widespread production of goods for commercial use. The advancement of biotechnology, modern engineering, and related advanced techniques has bridged the gap between traditional food fermentations and large-scale production methods. It ensures that product quality and safety are upheld (Schmidt, 2005).

In the past few years, there has been an increasing interest in combination of evolutionary engineering and other biotechnologies to achieve improved integration of microbial cells and enzymes. Traditional laboratory evolution is effective in promoting beneficial mutations in various genes, but it has certain limitations. These include a lengthy evolutionary process and unpredictable mutation rates. With advancements in genomics and synthetic biology have opened up new avenues for studying fermentation. These approaches, such as genomics, metabolomics, transcriptomics, metagenomics, and proteomics have allowed researchers to delve deeper into the process of fermentation by enabling the identification of markers and the enhancement of cellular transformation strategies. The generation of biomolecules with fermentation on a large scale has become more efficient, requiring less cost and time (Formenti et al., 2014).

Ongoing research and development in the process of fermentation is being conducted across various areas. This aims to enhance the quality and yield of products, while also reducing production costs and promoting environmentally friendly processes. To enhance the production of fermentation products involves fine-tuning the various factors that impact the process, including the microorganisms, the surrounding environment, and the technological infrastructure. Several factors have contributed to the optimization of the process:

- Feeding methods in bioreactors: batch, fed-batch, and continuous operation.
- Fermentation can involve either a single step or multiple steps, and it can occur in either solid or liquid form (Yang et al., 2018).
- Microbial cell types: single strain or mixed culture processes.
- Exploring the role of oxygen availability: aerobic, anaerobic, and microaerobic processes.
- Key factors in the process include the ratio of inoculation, the agitation rate for optimal mixing, and the continuous monitoring of pH levels (Peng et al., 2015).

Alternative methods integrate microorganisms and advanced technology at the micro-/nanoscale level. This case involves strategies that utilize electrochemistry, which has been widely recognized as effective method, particularly in treating wastewater and sludge.

1.5. APPLICATION OF FERMENTATION

Learning Objectives:

- Explore the application of fermentation in food and fuel.

1.5.1. Food Application

The significance of fermented products, such as fermented food products, lies in their ability to provide and preserve a wide range of nutritious foods with diverse textures, flavors, and aromas. This greatly enhances the overall quality of the food available for human consumption (Figure 1.5). Since the dawn of human history, fermented foods have existed on Earth (Goldrick et al.,

2019). It played a role in creating various food items, such as sausages, pickled vegetables, alcoholic beverages, yogurts, vinegar, cheese, sauces, and meat-flavored pastes. Along with the production of leavened and sourdough breads, their importance is expected to grow in the coming years. Developing countries have traditionally relied on natural biological systems, enriched using vitamins and essential amino acids, whereas developed countries have favored synthetic foods fortified with vitamins (Nedovic et al., 2011).

Figure 1.5. *Illustration of the various uses of fermented probiotics.*

Source: Faina Nakonechny, Creative Commons License.

Fermentation has seen significant advancements in various areas, including production of food items such as wine, cheese, bread, and beer and in the creation of high-value products, food chemicals, and medicines (Figure 1.6). The cost of petrol and the decline in fossil fuel availability have been significant factors contributing to the development of more cost-effective and simplified fermentation processes. Some research analysts have observed a significant surge in the academic and corporate sectors, driven by the increasing use of fermentation in various industries such as pharmaceuticals, energy, utilities, and special chemicals (Shahidi et al., 1999).

The field of biofuels has garnered increased attention due to the use of fermentation techniques to produce affordable bioethanol. Fermentation techniques have primarily found significance in the realm of fermented food. Fufu, Gari, Iru, and Ogi are examples of fermented foods. Irú is a kind of processed fermented food derived from locust beans (Parkia biglobosa) that have long been utilized as a condiment in various culinary applications. Fufu is commonly consumed as an everyday food in many African countries, including Ghana and Nigeria.

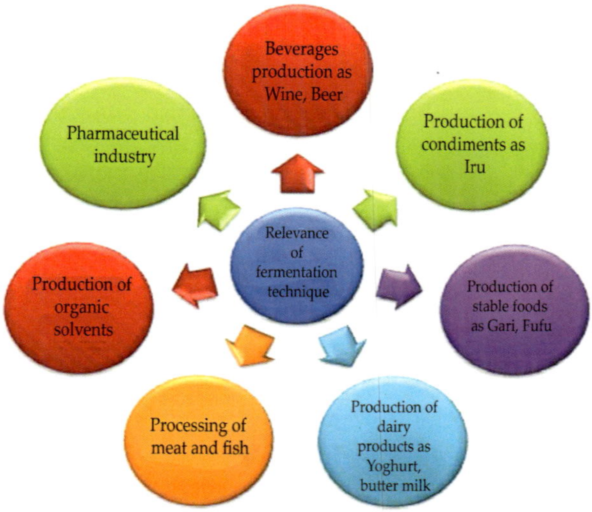

Figure 1.6. *Illustration of the relevance of fermentation technology in food processing.*

Source: Betty Niazov, Creative Commons License.

The mixture is prepared by combining equal amounts of cassava and green plantain flour with water. Gari is a finely textured flour derived from cassava tubers through fermentation process. It is widely consumed as a staple food (Hartmann & Meisel, 2007).

1.5.2. Fuel Application

1.5.2.1. Ethanol Fuel

Ethanol fermentation process is also known as alcohol fermentation, and represents a biological phenomenon that transforms sugars like fructose, glucose, and sucrose into cellular energy (Sen et al., 2016). Carbon dioxide and ethanol are thus created as byproducts. Anaerobic processes, such as alcoholic fermentation, are those in which yeasts convert alcohol without presence of oxygen. Besides lactic acid fermentation, several fish species including goldfish and carp also use fermentation as a low-oxygen mode of producing energy (Balat & Balat, 2009).

1.5.2.2. Hydrogen Gas Production in Fermentation

Fermentation in many forms produces hydrogen gas, which is further used to replace NAD+ in NADH. H_2 is produced when ferredoxin receives electrons and is oxidized by hydrogenase. Methanogens and sulfate reducers use hydrogen gas as a substrate, which helps to keep the concentration of hydrogen low and encourages the synthesis of an energy-dense molecule. Nonetheless, in a comparatively high quantity, hydrogen gas can develop, as in flatus (Bastidas et al., 2012).

One example of mixed acid fermentation involves the fermentation of glucose by bacteria like *Clostridium pasteurianum*. This process results in the production of butyrate, acetate, carbon dioxide, and hydrogen gas. The process resulting in the formation of acetate is:

$C_6H_{12}O_6 + 4\ H_2O \rightarrow 2\ CH_3COO^- + 2\ HCO_3^- + 4\ H^+ + 4\ H_2$

It is theoretically possible to convert glucose into only CO_2 and H_2, however, the overall reaction yields a minimal amount of energy (Ren et al., 2006).

1.5.2.3. Methane Gas Production in Fermentation

Another reaction that can occur with acetic acid is the dismutation reaction, resulting in the production of methane and carbon dioxide:

$CH_3COO^- + H^+ \rightarrow CH_4 + CO_2 \rightarrow \Delta G^\circ = -36$ kJ/reaction

Methanogen archaea catalyze this disproportionation reaction in their fermentative metabolism. An electron is transferred through the carbonyl function of a carboxylic group towards the methyl group of acetic acid, results in the production of CO_2 and methane gas (Lips, 2022).

1.6. FUTURE CHALLENGES OF FERMENTATION

Learning Objectives:

- Know the future challenges in fermentation technology and its applications.

As previously stated, fermentation plays a crucial role in various aspects of the production of food and beverages on a global scale. It also supports the production of a wide range of bio-compounds, including antibiotics, pigments, and bioplastics hormones. Despite the extensive research conducted on fermentation-based processes, which encompass a range of scientific disciplines including biochemistry, plant/microorganism genetics, biomass chemistry, and process engineering, the global utilization of fermentation still faces intriguing challenges shortly. This is especially important when bioethanol production used as a fuel alternative. It is worth noting that the process is still relatively slow compared to the high demand for biofuels globally (Chen & Liu, 2021).

In the current era of scientific advancements, one finds itself in a phase where numerous genomes of plants, fungi, and microorganisms involved in commercial fermentation or obtained through food fermentations have been successfully sequenced. This presents a fresh perspective on utilizing these microbes for fermentation in various industrial applications. It encompasses a range of activities, such as modifying microbes to create antimicrobials or nutritional through metabolic engineering, as well as exploring unknown molecular activities that could enhance food production and the development of bio-compounds for the market. Furthermore, the availability of the genomes of many pathogenic and spoilage bacteria may facilitate the creation of novel antibiotics that precisely target the vital functions of these bothersome bacteria. One of the key challenges in field of genomics and proteomics, particularly about food systems, is effectively utilizing the vast amount of information available to enhance culture performance and activities. Consequently, this could improve the composition, safety, and quality of the world's food supply (Walker, 2011).

Another significant aspect to consider is the role of technology. Certain microorganisms have been studied for industrial fermentations have been found to thrive in challenging environments, such as high salt concentrations or highly acidic pH levels. These microbes are halophilic or acid thermophilic in nature. Working in such conditions leads to corrosion impacts the lifespan of many bioreactors in use. As a result, the widespread use of these microorganisms is hindered (Rodriguez et al., 2014).

Designing fermentation processes focusing on a circular economy remains a significant challenge. Recent approaches have been focusing on utilizing food waste as raw materials to develop sustainable processes. These processes involve methanogenesis, solvent genesis, photosynthesis, acidogenesis, fermentation, oleaginous process, bioelectrogenesis, and more. To obtain a range of products such as biofuels, bioelectricity, platform chemicals, biomaterials, biofertilizers, animal feed, and more, which can be utilized for the valorization of fermentation waste (Jalili et al., 2023).

SUMMARY

- The chapter "Introduction to Fermentation" provides a foundational understanding of this vital biological process. Fermentation, defined as conversion of organic compounds into simpler substances by microorganisms or enzymes, plays a crucial role in various biological functions.
- The chapter outlines the biochemical mechanisms underpinning fermentation, emphasizing its significance in energy production and the synthesis of essential compounds. It delves into the principal types of fermentation.
- Microorganisms and enzymes are identified as key drivers of fermentation, facilitating the biochemical reactions necessary for product formation.
- Moreover, the chapter explores optimization of fermentation processes through technological advancements to improve efficiency and yield. It showcases the broad applications of fermentation in food and fuel production, underscoring its versatility and potential for sustainable solutions.
- However, the chapter also acknowledges future challenges facing fermentation technology, including scalability and resource utilization constraints.

REVIEW QUESTIONS

1. Define fermentation and explain its biological significance.
2. What are the principal types of fermentation discussed in the chapter? Provide brief explanations of each.
3. How do microorganisms and enzymes serve as key tools in fermentation processes?
4. Discuss the biochemical overview of fermentation provided in the chapter. What are the key mechanisms involved?
5. How can fermentation processes be optimized, and what is impact of technology on fermentation efficiency and yield?
6. Explore the applications of fermentation in both food and fuel production. What are some examples provided in the chapter?
7. What are future challenges facing fermentation technology, as discussed in the chapter? How might these challenges be addressed?

8. Can you identify any potential ethical or environmental considerations associated with the widespread application of fermentation technology?

MULTIPLE CHOICE QUESTIONS

1. What is the biological role of fermentation?

a. Synthesizing complex compounds

b. Breaking down inorganic compounds

c. Generating energy and producing simpler compounds

d. Facilitating photosynthesis

2. What are the primary tools utilized in fermentation processes?

a. Chemical solvents

b. Heat and pressure

c. Electricity and magnetism

d. Microorganisms and enzymes

3. What is the impact of technology on fermentation processes?

a. Decreases efficiency

b. Enhances efficiency and yield

c. Reduces yield

d. Slows down the process

4. In which industry is fermentation commonly applied?

a. Construction

b. Textile

c. Food

d. Automotive

5. What future challenges does fermentation technology face?

a. Decreasing complexity

b. Lowering costs

c. Addressing scalability issues and resource utilization constraints

d. Eliminating microbial involvement

6. Which of the following is a potential application of fermentation?

a. Petrochemical refining

b. Aerospace engineering

c. Waste management

d. Pharmaceutical manufacturing

Answers to Multiple Questions

1. (c); 2. (d); 3. (b); 4. (c); 5. (c); 6. (a)

REFERENCES

1. Aiba, S., Shoda, M., & Nagatani, M. (1968). Kinetics of product inhibition in alcohol fermentation. *Biotechnology and Bioengineering, 10*(6), 845–864.
2. Al-Maqtari, Q. A., Waleed, A. A., & Mahdi, A. A. (2019). Microbial enzymes produced by fermentation and their applications in the food industry: A review. *International Journal of Agriculture Innovations and Research, 8*(1), 2319–1473.
3. Arunkumar, R., Gorusupudi, A., & Bernstein, P. S. (2020). The macular carotenoids: A biochemical overview. *Biochimica et Biophysica Acta (BBA)–Molecular and Cell Biology of Lipids, 1865*(11), 158617.
4. Balat, M., & Balat, H. (2009). Recent trends in global production and utilization of bio-ethanol fuel. *Applied Energy, 86*(11), 2273–2282.
5. Bastidas-Oyanedel, J. R., Mohd-Zaki, Z., Zeng, R. J., Bernet, N., Pratt, S., Steyer, J. P., & Batstone, D. J. (2012). Gas controlled hydrogen fermentation. *Bioresource Technology, 110*, 503–509.
6. Bayrock, D., & Ingledew, W. M. (2001). Changes in steady state on introduction of a Lactobacillus contaminant to a continuous culture ethanol fermentation. *Journal of Industrial Microbiology and Biotechnology, 27*(1), 39–45.
7. Behera, S. S., & Ray, R. C. (2016). Solid state fermentation for production of microbial cellulases: Recent advances and improvement strategies. *International Journal of Biological Macromolecules, 86*(6), 656–669.
8. Chambers, R. W. (1985). Chemical carcinogenesis: A biochemical overview. *Clinical Biochemistry, 18*(3), 158–168.
9. Chen, G. Q., & Liu, X. (2021). On the future fermentation. *Microbial Biotechnology, 14*(1), 18–21.
10. Choi, H. K., Yoon, J. H., Kim, Y. S., & Kwon, D. Y. (2007). Metabolomic profiling of Cheonggukjang during fermentation by 1H NMR spectrometry and principal components analysis. *Process Biochemistry, 42*(2), 263–266.
11. Christenson, R. H. (1997). Biochemical markers of bone metabolism: An overview. *Clinical Biochemistry, 30*(8), 573–593.

12. Deville, Y., Gilbert, D., Van Helden, J., & Wodak, S. J. (2003). An overview of data models for the analysis of biochemical pathways. *Briefings in Bioinformatics, 4*(3), 246–259.
13. Dimidi, E., Cox, S. R., Rossi, M., & Whelan, K. (2019). Fermented foods: Definitions and characteristics, impact on the gut microbiota and effects on gastrointestinal health and disease. *Nutrients, 11*(8), 1806.
14. Florack, D. E. A., & Stiekema, W. J. (1994). Thionins: Properties, possible biological roles and mechanisms of action. *Plant Molecular Biology, 26*(1), 25–37.
15. Formenti, L. R., Nørregaard, A., Bolic, A., Hernandez, D. Q., Hagemann, T., Heins, A. L., & Gernaey, K. V. (2014). Challenges in industrial fermentation technology research. *Biotechnology Journal, 9*(6), 727–738.
16. Goldrick, S., Duran-Villalobos, C. A., Jankauskas, K., Lovett, D., Farid, S. S., & Lennox, B. (2019). Modern day monitoring and control challenges outlined on an industrial-scale benchmark fermentation process. *Computers & Chemical Engineering, 130*, 106471.
17. Gupta, R., Beg, Q., Khan, S., & Chauhan, B. (2002). An overview on fermentation, downstream processing and properties of microbial alkaline proteases. *Applied Microbiology and Biotechnology, 60*(3), 381–395.
18. Gurung, N., Ray, S., Bose, S., & Rai, V. (2013). A broader view: Microbial enzymes and their relevance in industries, medicine, and beyond. *BioMed Research International, 2013*(1), 329121.
19. Han, K., Lim, H. C., & Hong, J. (1992). Acetic acid formation in Escherichia coli fermentation. *Biotechnology and Bioengineering, 39*(6), 663–671.
20. Hartmann, R., & Meisel, H. (2007). Food-derived peptides with biological activity: From research to food applications. *Current Opinion in Biotechnology, 18*(2), 163–169.
21. Jalili, M., Nazari, M., & Magkos, F. (2023). Fermented foods in the management of obesity: Mechanisms of action and future challenges. *International Journal of Molecular Sciences, 24*(3), 2665.
22. John, R. P., Nampoothiri, K. M., & Pandey, A. (2007). Fermentative production of lactic acid from biomass: An overview on process developments and future perspectives. *Applied Microbiology and Biotechnology, 74*(6), 524–534.
23. Jørgensen, P., Pedersen, J. G., Jensen, E. P., & Esbensen, K. H. (2004). On-line batch fermentation process monitoring (NIR)—Introducing 'biological process time'. *Journal of Chemometrics, 18*(2), 81–91.

24. Kennedy, M., & Krouse, D. (1999). Strategies for improving fermentation medium performance: A review. *Journal of Industrial Microbiology and Biotechnology, 23*, 456–475.

25. Krishna, C. (2005). Solid-state fermentation systems—An overview. *Critical Reviews in Biotechnology, 25*(1–2), 1–30.

26. Lips, D. (2022). Fueling the future of sustainable sugar fermentation across generations. *Engineering Biology, 6*(1), 3–16.

27. Luo, H., Yang, R., Zhao, Y., Wang, Z., Liu, Z., Huang, M., & Zeng, Q. (2018). Recent advances and strategies in process and strain engineering for the production of butyric acid by microbial fermentation. *Bioresource Technology, 253*, 343–354.

28. Mayr, C. (2016). Evolution and biological roles of alternative 3′ UTRs. *Trends in Cell Biology, 26*(3), 227–237.

29. Montanari, G., Zambonelli, C., Grazia, L., Kamesheva, G. K., & Shigaeva, M. K. (1996). *Saccharomyces unisporus* as the principal alcoholic fermentation microorganism of traditional koumiss. *Journal of Dairy Research, 63*(2), 327–331.

30. Nedovic, V., Kalusevic, A., Manojlovic, V., Levic, S., & Bugarski, B. (2011). An overview of encapsulation technologies for food applications. *Procedia Food Science, 1*, 1806–1815.

31. Neubauer, P., Wolff, C., Hecker, M., Hofmann, K., Meyer, L., Kruschke, P., & Heinrich, H. W. (1991). Introduction of the tac-promoter by lactose under fermentation conditions. *Acta Biotechnological, 11*(1), 23–29.

32. Pan, Y., Wang, Y., Hao, W., Duan, C., Wang, S., Wei, J., & Liu, G. (2022). Metatranscriptomics unravel composition, drivers, and functions of the active microorganisms in light-flavor liquor fermentation. *Microbiology Spectrum, 10*(3), e02151–21.

33. Peng, B., Lei, Y., Zhao, H., & Cui, L. (2015). Response surface methodology for optimization of fermentation process parameters for improving apple wine quality. *Journal of Food Science and Technology, 52*, 7513–7518.

34. Reddy, G., Altaf, M. D., Naveena, B. J., Venkateshwar, M., & Kumar, E. V. (2008). Amylolytic bacterial lactic acid fermentation—A review. *Biotechnology Advances, 26*(1), 22–34.

35. Ren, N. Q., Li, Y. F., Wang, A. J., Li, J. Z., Ding, J., & Zadsar, M. (2006). Hydrogen production by fermentation: Review of a new approach to environmentally safe energy production. *Aquatic Ecosystem Health & Management, 9*(1), 39–42.

36. Rodriguez, B. A., Stowers, C. C., Pham, V., & Cox, B. M. (2014). The production of propionic acid, propanol and propylene via sugar fermentation: An industrial perspective on the progress, technical challenges and future outlook. *Green Chemistry, 16*(3), 1066–1076.
37. Schmidt, F. R. (2005). Optimization and scale up of industrial fermentation processes. *Applied Microbiology and Biotechnology, 68*, 425–435.
38. Sen, B., Aravind, J., Kanmani, P., & Lay, C. H. (2016). State of the art and future concept of food waste fermentation to bioenergy. *Renewable and Sustainable Energy Reviews, 53*, 547–557.
39. Shahidi, F., Arachchi, J. K. V., & Jeon, Y. J. (1999). Food applications of chitin and chitosans. *Trends in Food Science & Technology, 10*(2), 37–51.
40. Snyder, S. H., & Bredt, D. S. (1992). Biological roles of nitric oxide. *Scientific American, 266*(5), 68–77.
41. Valyasevi, R., & Rolle, R. S. (2002). An overview of small-scale food fermentation technologies in developing countries with special reference to Thailand: Scope for their improvement. *International Journal of Food Microbiology, 75*(3), 231–239.
42. Walker, G. M. (2011). 125th anniversary review: Fuel alcohol: Current production and future challenges. *Journal of the Institute of Brewing, 117*(1), 3–22.
43. Yang, X., Geng, B., Zhu, C., Li, H., He, B., & Guo, H. (2018). Fermentation performance optimization in an ectopic fermentation system. *Bioresource Technology, 260*, 329–337.

CHAPTER 2

CHEMISTRY AND FOOD FERMENTATION

LEARNING OBJECTIVES

At the end of this chapter, readers will be able to:

1. Understand primary and secondary metabolites and their roles in industrial microbiology.
2. Understand the significance and roles of fermentation in food processing.
3. Identify primary metabolites and their industrial applications, such as amino acids, organic acids and vitamins.
4. Describe the production processes and applications of secondary metabolites, including antibiotics, enzymes, and pigments.
5. Discuss the benefits associated with fermented foods, including their impact on gut microbiota and immune function.
6. Know the potential side effects, such as histamine toxicity from certain fermented products.

INTRODUCTORY EXAMPLE

Over the past few years, there has been a significant rise in the demand for fermented foods, as day by day people are looking for products that provide both taste and positive effects on their health. An interesting example of such a trend is the increasing popularity of kombucha, a tea beverage that undergoes fermentation and is recognized for its unique flavor and potential health benefits.

Established by passionate health-conscious individuals Sarah and Alex, Brewing Culture Co. set out on a mission to bring kombucha to a wider audience while staying true to its traditional origins. With a focus on precise fermentation techniques and creative flavor pairings, Brewing Culture Co. has rapidly gained popularity in the health food sector. They offer a diverse selection of kombucha options that are rich in probiotics and antioxidants. With every sip, customers not only indulge in a delightful drink but also experience the benefits of the fermentation process, showcasing Brewing Culture Co. as a remarkable illustration of the interconnectedness of chemistry and food fermentation.

UNIT INTRODUCTION

The United National Conference on Biological Diversity in 1992 provides a definition for biotechnology, which refers to the utilization of living creatures, biological systems, or their derivatives to produce or modify products or processes for specified objectives (Saisithi et al., 1966). Advancements in molecular biology have significantly enhanced our comprehension of genetics, leading to the emergence of modern biotechnology and its potential applications in recombinant DNA technology. This field, commonly known as genetic engineering, has profound impacts on the global economy. Advancements in molecular biology have paved the way for cutting-edge techniques like genetic engineering. This includes protoplast fusion, recombinant DNA technology, and cell culture. These methods allow in the modification of microbial cells, animals, and plants to enhance their biosynthesis capabilities. These modern biotechnologies enables production of valuable products through various biological processes (Ojha et al., 2017).

The main directions for growth in modern biotechnology are influenced by recent breakthroughs in molecular biology, as well as the pressing concerns of society regarding issues such as the energy crisis, food scarcity, ecological challenges, and bioremediation.

Industrial microbiology has grown as a modern scientific field, building upon advances in general microbiology, biochemistry, and biotechnologies.

Throughout history, microorganisms have been utilized for various purposes such as wine production, food preservation, and in creation of fermented milk and cheese. Before the advent of industrial cooling technology, different methods were employed to preserve food. These included acidification through lactic or acetic fermentation, and alcoholic fermentation, desiccation, and the addition of sugar to raise osmotic pressure (Admassie, 2018).

The bio-industry, often known as biotechnology, focuses on utilizing the industrial potential of microbes, plant or animal cells, and their subcellular components to acquire useful goods. These processes, which encompass biosynthesis or bioconversion, are executed by microorganisms with substantial economic significance. The name which is composed of two terms: bios and method, which denotes a manufacturing process involving living cells. Various aspects of contemporary biotechnology have greatly influenced conventional biotechnology (Mani, 2018).

In this chapter, we delve into the fascinating relationship between chemistry and food fermentation, discussing a range of important subjects. Exploring the realm of biotechnological products, the fascinating world of primary and secondary metabolites is explored. An in-depth analysis is conducted on the fermentation process, highlighting the role of microorganisms in the conversion of food substances. The chapter thoroughly examines the effects of consuming food that is fermented, including the positive impact on digestion and potential drawbacks. It specifically highlights the benefits of fermented dairy products. In addition, it delves into various fermented foods and associated health advantages, by thorough examination of this age-old tradition and its contemporary uses (Sharma et al., 2020).

2.1. PRIMARY AND SECONDARY METABOLITES AS BIOTECHNOLOGICAL PRODUCTS

Learning Objectives:

- Understand the distinction between primary and secondary metabolites in biotechnology.
- Explore the industrial applications and production methods of primary and secondary metabolites.

In industrial microbiological processes, production is achieved through the metabolic activity of microorganisms found within or on the transformed product. It was noted by researchers, when microorganisms that were not carefully chosen were used, the results were not consistent and led to significant variations in the yield. Furthermore, the presence of spontaneous microbiota can be accompanied by impurities can have an impact on the overall quality of the final product. To avoid these drawbacks, industrial strains are carefully chosen for their metabolic properties, resulting in a consistently superior product (Verpoorte et al., 2002).

The process of metabolism encompasses all the biochemical reactions performed by an organism. There are two main paths involved in this process: primary metabolic pathways and secondary metabolic pathways. Understanding the distinction among these pathways is of utmost importance in the field of industrial microbiology, even though the exact boundary between them may not be clearly defined. The use of nutrients within the cell's environment, particularly those with small molecules, is essential for supporting cellular functions through primary metabolic pathways. Primary metabolism can occur through three possible mechanisms – the hexose monophosphate pathway (HMP), Embden Meyerhof–Parnas pathway (EMP), and the Entner–Dourdorof pathway are three different metabolic pathways (Croteau et al., 2000).

By manipulating metabolic regulatory mechanisms, it is possible to enhance the creation of both primary and secondary metabolites in microbial organisms. This can significantly improves the industrial performance of specific microbial strains. Various factors and molecules are involved in metabolite production, affecting positive and negative mechanisms. These include inducers, effectors, inhibitors, and signal molecules, which contribute to the overproduction of various metabolite types. The regulation mechanisms include induction, nutritional factors, and feedback regulation (Pagare et al., 2015).

2.1.1. Primary Metabolites

Microbial products known as primary metabolites are produced through the exponential stages of growth (*see* Figure 2.1). The synthesis of these products is a natural part of the growth process. Primary metabolites exhibit a wide range of sizes, spanning from hydrogen gas, which weighs 2 Da, to vitamin B12, which weighs 1,355 Da. Key components in the industrial sector include solvents, organic acids, vitamins, amino acids, and nucleotides. Additionally, fermentation yields important products from primary metabolism, including ethanol, lactic acid, and acetic acid. These chemicals are essential components of the fermentation industry and have several industrial applications. These products are created through the action of a wide variety of bacteria and fungi with multiple applications in the food, chemical, and nutraceutical industries (Drew & Demain, 1977).

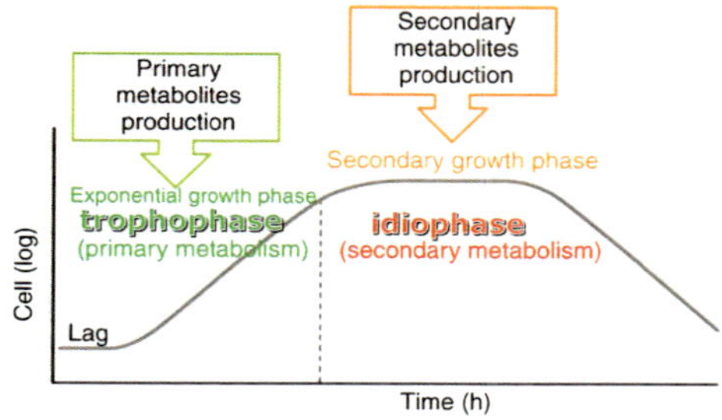

Figure 2.1. *Representation of growth curve and metabolite production in microorganisms.*

Source: Carmen Socaciu, Creative Commons License.

Many of such metabolites are produced through microbial fermentation instead of chemical synthesis because of fermentation processes are cost-effective and result in biologically valuable isomeric forms. Various industrially significant chemicals can be produced through microbial fermentations. For instance, glycerol and other polyhydroxy alcohols fall into this category. However, at the moment, these chemicals can be made inexpensively as by-products of petroleum. The recent rise in petroleum costs has revived interest in utilizing microbes for production of organic acids, solvents, and ethanol (Erb & Kliebenstein, 2020).

In the realm of microbiology in industry, alcohol stands out as a prevalent primary metabolite employed for large-scale production. In certain processes, alcohol is utilized for fermentation, results in creation of various products like beer and wine (Bolton, 2009).

Typical supplements include amino acids like l-glutamate and l-lysine, which are primary metabolites. In the realm of industrial processes, *Corynebacterium glutamicum* strains are used, with molasses serving as their growth media. This molasses is derived from starch hydrolysis, resulting in the presence of glucose, fructose, and sucrose. Commercially, L-Amino acids find wide application as additives in various industries such as therapeutic agents, feed supplements, food, infusion compounds, or predecessors for peptide production or agriculture-based chemicals (Molina et al., 2006).

Another instance of a primary metabolite frequently employed in industrial microbiology is citric acid. The synthesis of citric acid, a commonly utilized commercial product, is achieved through the largest fermenting process. This process mainly involves the use of carefully selected fungal strains (such as *Aspergillus niger*) or strains of yeast in aerobe bioreactors. To decrease expenses and address environmental concerns, the byproducts of the agricultural industry such as molasses and glycerin are used as substrates in the manufacturing of citric acid. Continuous processes may be used as an alternative to traditional fungi methods to achieve higher yields of citric acid. In this study, a mutant strain of *A. niger*, called H4002, was developed via accelerated carbon ions and X-ray methods to enhance citric acid production. This resulted in citric acid levels reaching 200–250 g/L. Under carefully controlled culture conditions, the H4002 strains showed impressive results, producing 187.5 ± 0.7 g/L of citric acid. This highlights their potential for usage in the industrial synthesis of citric acid (Craney et al., 2012). Citric acid is widely utilized in various industries, including cosmetics, pharmaceuticals, food, beverage and chemicals. It serves multiple purposes such as enhancement, acidulation, antioxidant, plasticization, and flavor preservation, acting as a synergistic agent. At a dose of 1–2 g/kg when taken orally, citric acid has been shown to have antioxidant and anti-inflammatory properties in brain tissue. Additionally, it has been found to have a protective effect on the liver within this dosage range (Asai et al., 2017).

Essential micronutrients, including vitamins (thiamine, riboflavin, folic acid, biotin, and pantothenic acid), playing a crucial role in various enzymatic pathways. These micronutrients are synthesized exclusively by microorganisms like yeasts and bacteria, as well as plants, and act as cofactors. Microorganisms can produce vitamins cost-effectively, with numerous species having undergone genetic engineering to increase the production of these molecules. Vitamins and related chemicals are presently generatedusing fermentation methods and microbial or enzymatic transformation. These substances have found extensive applications as additives in food and feed, as well as in the fields of medicine, therapy, health, cosmetics, and technology (Schwachtje & Baldwin, 2008).

2.1.2. Secondary Metabolites

Microorganisms produce secondary metabolites, which are small biomolecules are not necessary for the generating organism to survive during the stationary phase. The substances encompass enzymes, antibiotics, antitumor agents, and antivirals, which have extensive applications in the field of medicine. Additionally, phytotoxins or herbicides utilized in agriculture, as well as pigments and flavors employed as food additives (Vining, 1990).

Did You Know?

Fermented foods, such as kimchi and sauerkraut, offer more than just delicious flavors. And can play a role in supporting weight management through their positive effects on digestion and the promotion of a healthy gut microbiome.

There has been a substantial amount of advancement in the field of genomics, notably in the investigation of new secondary metabolites, which is referred to as metabolomics. Metabolomics play a crucial role in synthetic biology, as it focuses in the exploration and analysis of novel secondary metabolites. Many microorganisms possess the genetic potential to produce a larger variety of secondary metabolites than initially anticipated (Guerriero et al., 2018). Through bioinformatics evaluation of putative secondary metabolism gene clusters within sequenced genomes, it is possible to anticipate the possibility of cryptic secondary metabolites. This is because the gene expression of these metabolites is typically low when the levels of the final products are undetected. The primary objective of the most recent research is to explore innovative techniques that can stimulate the expression of hidden metabolic genes and discover previously unknown compounds (Bennett & Wallsgrove, 1994).

Microorganisms are a valuable source of enzymes, particularly in realm of industrial bioprocesses. To generate significant amounts of enzymes, it is possible to culture microorganism strains rapidly through fermentation. These strains can be readily modified at the genetic level, enabling the synthesis of industrial enzymes using genetically altered bacteria and fungi (Pagare et al., 2015).

Microbial enzymes are utilized in multitude of industries, spanning from paper and pulp to leather, detergents, food and beverages, textiles, pharmaceuticals, animal feed, biofuels, chemicals, and cosmetics. The study of lipolytic enzymes has garnered considerable attention due to their immense biotechnological promise. Lipases have high impact on various industries including food, chemicals, detergent, and pharmaceuticals and are considered valuable enzymes in biotechnology due to their versatile properties and wide range of applications. One of the functions of amylase is to help break down starch in sugars, results in the creation of glucose and fructose syrup. The role of α-Amylase catalysis is crucial in this process, with wide-ranging applications in production of fructose syrup, eco-friendly detergents, and baked goods. Certain microorganisms, including bacteria, fungi, and genetically modified species, can produce α-Amylase (Croteau et al., 2000).

Microbial amylase is highly favored due to its versatility and wide availability compared to other sources. Two commonly utilized bacterial species are Bacillus amyloliquefaciens and Bacillus licheniformis (Tiwari & Rana, 2015).

Penicillin, discovered by Fleming in 1928, has been the focus of much research, establishing antibiotics as highly studied secondary metabolites. The antibiosis procedure is a competitive strategy used by soil microbial strains, including Actinobacteria, Myxobacteria, fungi, and Cyanobacteria, to survive in natural environments. This process is a major source of antibiotic synthesis. Each species in the environment produces several antibiotics, and the composition of these antibiotics is distinct to each species. Recent advancements in scientific techniques and equipment have revealed that microorganisms can produce wide range of antibiotics. Within natural microbial communities, *Streptomyces* spp. stands out as the dominant genus responsible for antibiotic production. However, this microbe specifically produces antibacterial molecules that are structurally related (Vaishnav & Demain, 2011).

Modern methods in scientific research focus on identifying new compounds by studying particular and lesser-known microbial populations found in distinct environments. These environments include specific areas on land, the relationships between plants and microorganisms, and the vast marine ecosystems (Edreva et al., 2008).

The distribution of certain microbial species, even within common taxa, is influenced by environmental conditions. These conditions act as strong selection factors and can result in the production of unique compounds. There has been a growing focus among research groups on investigating previously unexplored microbial communities found in various environments such as rhizospheres, lichens, endolithic microbial communities, marine sediments, endosymbionts, insect parasites, plant endophytes, and invertebrates. These methods have been beneficial in identifying unique microbial communities that may produce new chemical structures (Verma & Shukla, 2015).

2.2. FERMENTATION PROCESS

Learning Objectives:

- Understand the biochemical principles and diverse roles of fermentation in food processing.

Food fermentation is an ancient technique used in transforming food into a more durable state, ensuring its long-term preservation. The word comes from the Latin for fever (to boil), which describes the chemical change in microbes and their byproducts create in a fermenter that frequently results in bubbling and warmth. Louis Pasteur (1822–1895) carried out pioneering experiments that unveiled the process of fermentation as a type of anaerobic metabolism, involving living yeast. His work demonstrated the transformation of glucose into ethanol, explaining fermentation is a form of respiration in the absence of air. Pasteur's research provided valuable insights into the complex phenomenon of fermentation (Schäpper et al., 2009).

Practice Problem

K&L Food a famous food production company aims to diversify its product line by introducing fermented foods. As a consultant, outline the potential benefits and challenges of incorporating fermented foods into their product offerings. Provide recommendations for mitigating any identified challenges.

Solutions to Practice Problem

Incorporating fermented foods into K&L Food Company's product line presents numerous benefits, including enhanced flavor profiles, extended shelf life, and increased nutritional value due to the presence of probiotics. However, challenges such as maintaining consistent quality, managing production costs, and navigating regulatory requirements may arise. To mitigate these challenges, K&L Food Company should invest in rigorous quality control measures, streamline production processes to optimize efficiency, and collaborate with regulatory experts to ensure compliance. Additionally, fostering transparent communication with consumers about the health benefits and unique characteristics of fermented foods can build trust and stimulate demand. By strategically addressing these challenges, K&L Food Company can successfully diversify its product line and capitalize on the growing popularity of fermented foods.

According to Walker, fermentation is a procedure where microorganisms or enzymes derived from plants or animals break down organic substances over time (United Nations Food and Agriculture Organization, FAO) (Schmidt, 2005).

During the process of fermentation, specific organic compounds, typically two different metabolites, obtained from the fermentable substrate, act as electron donors and acceptors, respectively (Figure 2.2). The breakdown of fermentative substrates is not fully efficient, resulting in a significantly lower energy release compared to respiration, where the oxidation process is more thorough. The initiation of fermentation processes occurs through phosphorylation on the substrate level. The outcome of their research is the production of ATP and various other compounds that possess high-energy bonds, with the most significant one being acetyl-CoA (Vandamme & Derycke, 1983).

Fermentation is a crucial process within the food industry, serving multiple roles in food processing:

- To enhance the nutritional value of the human diet by creating a broad range of flavors, smells, and textures in food.

- The preservation of significant quantities of food is achieved using alcoholic, alkaline fermentations, acetic acid, lactic acid, and high salt fermentations (Konstantinov & Yoshida, 1989).
- To enhance the nutritional content of dietary substrates through the addition of essential vitamins, protein, amino acids, and essential fatty acids.
- To explore the process of detoxification through food fermentation.
- To reduce cooking durations and energy consumption (Li et al., 2011).

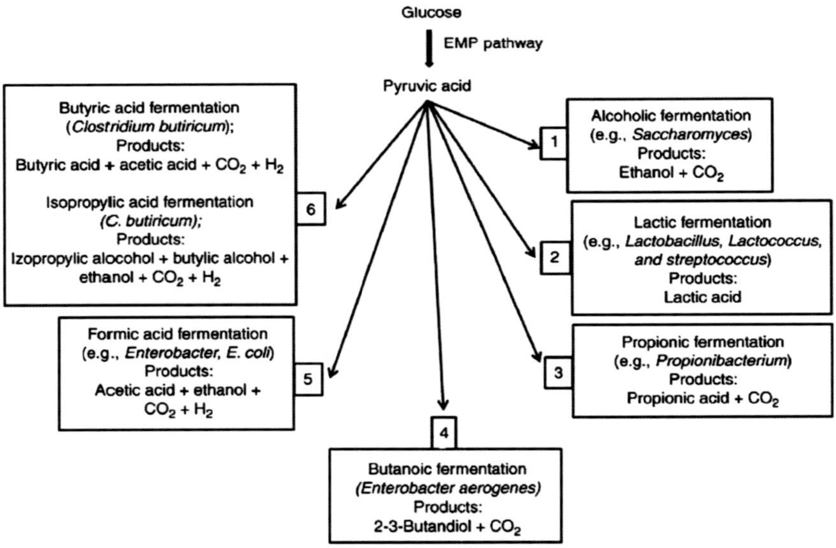

Figure 2.2. *Schematic illustration of the fermentative process based on the final products.*

Source: Henryk Jelen, Creative Commons License.

2.3. FERMENTATIVE MICROORGANISMS

Learning Objectives:

- Comprehend the selection criteria and characteristics of fermentative microorganisms in industry.

Fermentation is an organic process that occurs due to microorganisms and is utilized by humans on a large scale, although not created by humans. In food fermentation, there has been significant progress in the identification and categorization of microorganisms. A comparison between the "2002 IDF Inventory" and the current "Inventory of MFC" reveals a substantial increase in bacterial species and species of yeasts and molds. These microorganisms play a crucial role in food fermentations, spanning various food matrices like fish, meat, beverages, vegetables, dairy, and vinegar. Industrial microbiology utilizes a wide range of microorganisms, including naturally occurring organisms, laboratory selected mutants, and genetically modified organisms, to produce a diverse array of industrial products on a large scale. When it comes to biotechnology processes, it is essential to utilize microorganisms considered safe by the Food and Drug Administration (FDA). This is especially important in the manufacturing of food products and ingredients. For a substance or microorganism to be considered safe, it must have received approval from an expert who is qualified to assess its safety (Gaden, 1959).

Did You Know?

Fish sauce, a common ingredient in Southeast Asian dishes, has been linked to potential health concerns such as esophageal and gastric cancers. This emphasizes the need for moderation when consuming fermented foods.

Understanding the unique characteristics of microorganisms enables their practical use in a wide range of industries, including agriculture, food production, pharmaceuticals, chemicals, leather, and environmental conservation. After the selection of a microorganism, whether through traditional methods or modern molecular techniques, it becomes essential to grow it under specific conditions (Hartmann et al., 2015). These conditions guarantee the manifestation of particular traits that prove beneficial in industrial applications. Valuable substances can be characterized by the components of microbial cells or the substances released into the surrounding atmosphere through the growth of microbes. There are different types of microorganisms utilized in industrial microbiology, including bacteria, microscopic fungi, such as yeasts and molds, and microalgae (Schmidt, 2005) (Table 2.1).

To be utilized in the commercial microbiology field, microorganisms must meet certain criteria (Figure 2.3):

- The substance of interest must produceby the industrial microorganism.
- It is essential to acquire microbes of interest in a pure culture.
- The organism must maintain genetic stability and be easily modifiable through genetic manipulation (Pandey, 1992).
- Requires cultivation in extensive laboratory cultures.

- It is essential to sustain pure cultures of microorganisms for an extended duration under controlled laboratory conditions.
- The microorganisms should exhibit in strong development once introduced into the seed-stage vessel (Büscher et al., 2009).
- The desired product should be produced within very short amount of time.
- The microorganism should be capable of thriving in a cost-effective liquid culture medium that can be easily obtained in large quantities.
- The production process should aim to create a desirable end product that is easily obtained, with a preference for a single product and without presence of any harmful by-products.
- An industrial microorganism must pose no harm to humans or economically significant animals and plants (Caplice & Fitzgerald, 1999).
- The microorganism should be able to defend itself against contamination, such as by producing antimicrobial substances. Additionally, it should be resistant to bacteriophages to ensure its stability.
- In industrial settings, microorganisms with larger cell sizes are highly preferred. This is due to larger cells have the advantage of settling quickly from a culture and can be efficiently filtered out using cost-effective filter materials (Parker et al., 2018).

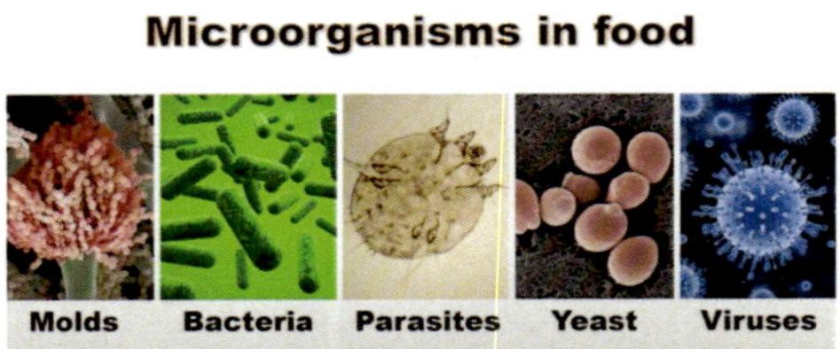

Figure 2.3. *Illustration of the microorganisms in food.*

Source: Amelia Stowell, Creative Commons License.

Table 2.1. *The main biotechnological microorganisms and their products (Source: Henryk Jelen, Creative Commons License)*

Application Fields	Microorganisms	Types	Products
Food and beverage industry	Saccharomyces cerevisiae	Yeast	Bakery, wine, beer, saké
	Saccharomyces carlsbergensis	Yeast	Light beer

	Saccharomyces rouxii	Yeast	Soy sauce
	Candida milleri	Yeast	Sour dough French bread (sour bread)
	Lactobacillus san-francisco	Bacteria	Sour dough French bread ("sour bread")
	Streptococcus ther-mophilus	Bacteria	Yogurt
	Lactobacillus bul-garicus	Bacteria	Yogurt
	Propionibacterium shermanii	Bacteria	Swiss cheese
	Gluconobacter suboxydans	Bacteria	Vinegar
	Penicillium roque-forti	Filamentous fungi	"Roquefort" cheese
	Penicillium camem-berti	Filamentous fungi	"Camembert" and "Brie" cheeses
	Aspergillus oryzae	Filamentous fungi	Saké
	Rhizopus sp.	Filamentous fungi	Tempeh
	Mucor sp.	Filamentous fungi	Sufu (microbial rennet)
	Monascus purpu-reus	Filamentous fungi	Ang-kak (red rice)
Chemical industry	S. cerevisiae	Yeast	Ethanol (from glucose)
	Kluyveromyces fragilis	Yeast	Ethanol (from glucose)
	Clostridium aceto-butylicum	Bacteria	Acetone and butanol
	Xanthomonas campestris	Bacteria	Polysaccharides
	A. niger	Filamentous fungi	Citric acid

Aromatic amino acids and nucleotides	C. glutamicum	Bacteria	L-Lysine, 5¢-inosinic acid, 5¢-guanylic acid
SCP	Candida utilis	Yeast	SCP from paper industry waste cultivation
	Yarrowia lipolytica	Yeast	SCP from alkane cultivation
	Methylophilus methylotrophus	Bacteria	SCP from methane or methanol cultivation
Vitamins	Eremothecium ashbyi	Yeast	Riboflavin
	Pseudomonas denitrificans	Bacteria	Vitamin B12
	Propionibacterium	Bacteria	Vitamin B12
Enzymes	Aspergillus oryzae	Filamentous fungi	Amylase
	A. niger	Filamentous fungi	Glucoamylase
	Trichoderma reesei	Filamentous fungi	Cellulase
	S. cerevisiae	Yeast	Invertase
	K. fragilis	Yeast	Lactase
	S. lipolytica	Yeast	Lipase
	Bacillus	Yeast	Protease
	Aspergillus	Yeast	Pectinase and protease
	Endothia parasitica	Yeast	Microbial rennet
Polysaccharides	Leuconostoc mesenteroides	Bacteria	Dextran
	X. campestris	Bacteria	Xanthan gum
Carotenoids	Blakeslea trispora	Filamentous fungi	β-Carotene
	Phaffia rhodozyma	Yeast	Astaxantine
Pharmaceutical industry	Penicillium chrysogenum	Filamentous fungi	Penicillin

2.4. BENEFITS AND SIDE EFFECTS OF FERMENTED FOODS

Learning Objectives:

- Understand health benefits and potential risks associated with fermented foods.

2.4.1. Fermented Dairy Products and Their Benefits

Several studies demonstrated the positive impact of fermented dairy products on human health, including their potential to reduce the risk of cancer, cardiovascular diseases, and periodontitis. Fermented dairy products have been found to positively impact immune and metabolic systems. Which not only provide beneficial effects but also help stimulate the growth of healthy bacteria in the intestines when consumed with probiotics. Additionally, these products act as antibodies, reducing the chance of respiratory infections (Buttriss, 1997).

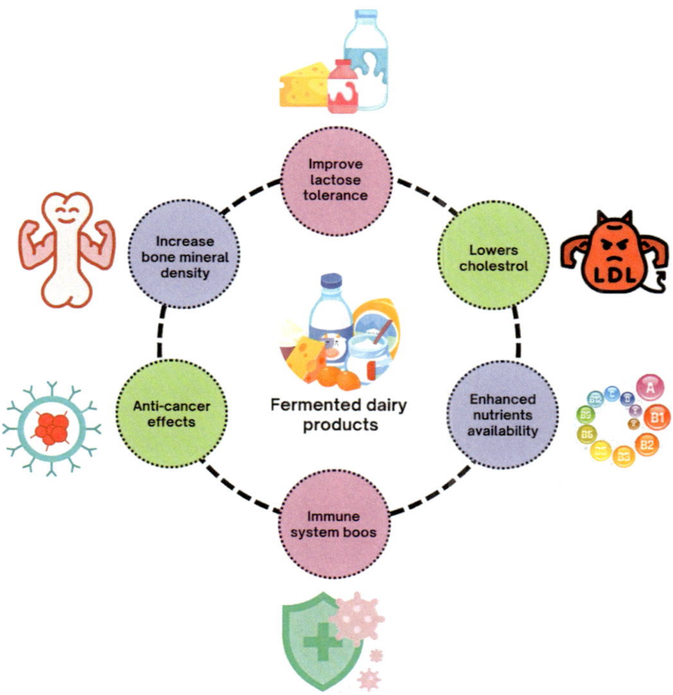

Figure 2.4. *Illustration of the therapeutic effects of fermented dairy products (probiotics).*

Source: Clarisse Ruth Betsabe Cuvas, Creative Commons License.

There is an abundance of research demonstrating the positive effects of probiotic food on immune-compromised patients. These studies have explored ina range of areas including transplantation, skin infections, respiratory infections, abdominal surgery, decreasing pathogenic microbes within the nasal passages, arthritis, and diabetes treatment (Agyei et al., 2020) (Figure 2.4).

2.4.2. Additional Fermented Foods and Their Health Benefits

Kimchi, a fermented vegetable, offers a wide range of health benefits. It has been associated with potential cancer prevention, weight management, immune system support, cholesterol reduction, and digestive health improvement. The apple is a fruit that offers numerous health benefits thanks to its rich polyphenol content. These antioxidants help lower the risk of chronic illnesses like cardiovascular disease, making them a valuable addition to a healthy diet. Fermented apple pomace is derived from the solid-state fermentation process of apple pomace. The concentration of yeast and the nutritional value of the protein both increase when pomace undergoes fermentation, due to the growth of microorganisms (Şanlier et al., 2019).

Fermented apple pomace can preserve inherent qualities of fruits, enhance the fermentation within the rumen, and contribute to increased milk production (Rul et al., 2022).

2.4.3. Fermented Foods Side Effects

A common element in the food of many Southeast Asian civilizations is fish sauce. However, there is a higher risk of stomach and esophageal squamous cell carcinomas if this fermented sauce is regularly consumed. Various scientific studies have indicated that the increased occurrence of tumor diseases can be attributed in the presence of N-nitroso compounds and genotoxins in fermented fish products (Baruah et al., 2022). Furthermore, the presence of biogenic amines in certain foods, produced by specific bacteria such as *Enterobacteriaceae*, having significant effects on both human health and the overall quality of the food. In various food products, such as fermented fish, histamine acts as a biogenic amine which can be found in varying concentrations. Individuals who have low amine oxidase activity, whether due to genetic predisposition, gastrointestinal diseases, or medication with monoamine oxidase inhibitors, may be susceptible to histamine toxicity. In contrast, those with normal amine oxidase activity can efficiently detoxify dietary histamine (Bell et al., 2018).

SUMMARY

- The chapter explores the symbiotic relationship between chemistry and food fermentation. It discusses the production of biotechnological products through primary and secondary metabolites, detailing their roles and applications.
- The fermentation process, driven by fermentative microorganisms, are explained, emphasizing its significance in food preservation and flavor development.
- The health benefits of fermented foods, including improved digestion and enhanced nutrient absorption, are outlined, alongside potential side effects such as allergic reactions. Specific focus is given to fermented dairy products and their benefits, as well as other fermented foods and their diverse health advantages.
- Lastly, the chapter addresses the potential side effects of fermented foods, emphasizing the importance of moderation and individual tolerance.

REVIEW QUESTIONS

1. Explain the difference between primary and secondary metabolites in biotechnological products.
2. How do fermentative microorganisms contribute to food fermentation?
3. Discuss the health benefits of consuming fermented foods, with examples.
4. What are the potential side effects of fermented foods, and how can they be managed?
5. Explain the significance of fermented dairy products for gut health and lactose digestion.
6. Explore examples of fermented foods beyond dairy and their health benefits.
7. How can individuals incorporate fermented foods into their diets while considering moderation and tolerance?
8. Discuss the role of fermentation in food preservation and flavor development, with examples.
9. How do primary and secondary metabolites contribute to fermentation biotechnology? Provide examples.

MULTIPLE CHOICE QUESTIONS

1. **Which category of metabolites is primarily essential for growth and reproduction?**
   a. Primary metabolites
   b. Secondary metabolites
   c. Tertiary metabolites
   d. Quaternary metabolites
2. **What is the main role of fermentative microorganisms in the fermentation process?**
   a. Enhancing food color
   b. Producing toxins
   c. Facilitating food preservation
   d. Reducing nutrient content
3. **What are the health benefits associated with consuming fermented foods?**
   a. Increased risk of allergies
   b. Improved digestion
   c. Decreased nutrient absorption
   d. Heightened cholesterol levels
4. **Which type of fermented food is known for its probiotic content and improved lactose digestion?**
   a. Fermented vegetables
   b. Fermented fruits
   c. Fermented dairy products
   d. Fermented meats
5. **What is emphasized regarding the consumption of fermented foods?**
   a. Unlimited consumption for optimal health benefits
   b. Potential side effects should be ignored
   c. Moderation and individual tolerance are important
   d. Fermented foods are not suitable for any diet

Answers to Multiple Questions

1. (a); 2. (c); 3. (b); 4. (c); 5. (c)

REFERENCES

1. Admassie, M. (2018). A review on food fermentation and the biotechnology of lactic acid bacteria. *World Journal of Food Science and Technology, 2*(1), 19–24.
2. Agyei, D., Owusu-Kwarteng, J., Akabanda, F., & Akomea-Frempong, S. (2020). Indigenous African fermented dairy products: Processing technology, microbiology and health benefits. *Critical Reviews in Food Science and Nutrition, 60*(6), 991–1006.
3. Asai, T., Matsukawa, T., & Kajiyama, S. I. (2017). Metabolomic analysis of primary metabolites in citrus leaf during defense responses. *Journal of Bioscience and Bioengineering, 123*(3), 376–381.
4. Asghar, F., Ali, S., Goraya, A., Javaid, I., & Hussain, Z. (2017). A review on the role of fermented foods as health promoters. *International Journal of Scientific Research in Science and Technology, 3*(1), 141–148.
5. Baruah, R., Ray, M., & Halami, P. M. (2022). Preventive and therapeutic aspects of fermented foods. *Journal of Applied Microbiology, 132*(5), 3476–3489.
6. Beena Divya, J., Kulangara Varsha, K., Madhavan Nampoothiri, K., Ismail, B., & Pandey, A. (2012). Probiotic fermented foods for health benefits. *Engineering in Life Sciences, 12*(4), 377–390.
7. Bell, V., Ferrão, J., Pimentel, L., Pintado, M., & Fernandes, T. (2018). One health, fermented foods, and gut microbiota. *Foods, 7*(12), 195.
8. Bennett, R. N., & Wallsgrove, R. M. (1994). Secondary metabolites in plant defense mechanisms. *New Phytologist, 127*(4), 617–633.
9. Bolton, M. D. (2009). Primary metabolism and plant defense—Fuel for the fire. *Molecular Plant-Microbe Interactions, 22*(5), 487–497.
10. Büscher, J. M., Czernik, D., Ewald, J. C., Sauer, U., & Zamboni, N. (2009). Cross-platform comparison of methods for quantitative metabolomics of primary metabolism. *Analytical Chemistry, 81*(6), 2135–2143.
11. Buttriss, J. (1997). Nutritional properties of fermented milk products. *International Journal of Dairy Technology, 50*(1), 21–27.
12. Caplice, E., & Fitzgerald, G. F. (1999). Food fermentations: Role of microorganisms in food production and preservation. *International Journal of Food Microbiology, 50*(1–2), 131–149.
13. Craney, A., Ozimok, C., Pimentel-Elardo, S. M., Capretta, A., & Nodwell, J. R. (2012). Chemical perturbation of secondary metabolism demonstrates important links to primary metabolism. *Chemistry & Biology, 19*(8), 1020–1027.

14. Croteau, R., Kutchan, T. M., & Lewis, N. G. (2000). Natural products (secondary metabolites). *Biochemistry and Molecular Biology of Plants, 24*(5), 1250–1319.
15. De Jager, C. M., Butôt, R. P. T., Van Der Meijden, E., & Verpoorte, R. (1996). The role of primary and secondary metabolites in chrysanthemum resistance to *Frankliniella occidentalis*. *22*(7), 1987–1999.
16. Drew, S. W., & Demain, A. L. (1977). Effect of primary metabolites on secondary metabolism. *Annual Review of Microbiology, 31*(1), 343–356.
17. Edreva, A., Velikova, V., Tsonev, T., Dagnon, S., Gürel, A., Aktaş, L., & Gesheva, E. (2008). Stress-protective role of secondary metabolites: Diversity of functions and mechanisms. *Genetics and Applied Plant Physiology, 34*(1–2), 67–78.
18. Erb, M., & Kliebenstein, D. J. (2020). Plant secondary metabolites as defenses, regulators, and primary metabolites: The blurred functional trichotomy. *Plant Physiology, 184*(1), 39–52.
19. Gaden Jr, E. L. (1959). Fermentation process kinetics. *Journal of Biochemical and Microbiological Technology and Engineering, 1*(4), 413–429.
20. Guerriero, G., Berni, R., Muñoz-Sanchez, J. A., Apone, F., Abdel-Salam, E. M., Qahtan, A. A., & Faisal, M. (2018). Production of plant secondary metabolites. *Genes, 9*(6), 309.
21. Hartmann, A., Albert, A., & Ganzera, M. (2015). Effects of elevated ultraviolet radiation on primary metabolites in selected alpine algae and cyanobacteria. *149*, 149–155.
22. Konstantinov, K., & Yoshida, T. (1989). Physiological state control of fermentation processes. *Biotechnology and Bioengineering, 33*(9), 1145–1156.
23. Li, X. R., Ma, E. B., Yan, L. Z., Meng, H., Du, X. W., Zhang, S. W., & Quan, Z. X. (2011). Bacterial and fungal diversity in the traditional Chinese liquor fermentation process. *International Journal of Food Microbiology, 146*(1), 31–37.
24. Mani, A. (2018). Food preservation by fermentation and fermented food products. *International Journal of Academic Research and Development, 1*, 51–57.
25. Marco, M. L., Heeney, D., Binda, S., Cifelli, C. J., Cotter, P. D., Foligné, B., & Hutkins, R. (2017). Health benefits of fermented foods: Microbiota and beyond. *Current Opinion in Biotechnology, 44*, 94–102.
26. Molina-Molina, J. M., Hillenweck, A., Jouanin, I., Zalko, D., Cravedi,

J. P., Fernández, M. F., & Balaguer, P. (2006). Steroid receptor profiling of vinclozolin and its primary metabolites. *Toxicology and Applied Pharmacology, 216*(1), 44–54.

27. Ojha, K. S., Mason, T. J., O'Donnell, C. P., Kerry, J. P., & Tiwari, B. K. (2017). Ultrasound technology for food fermentation applications. *Ultrasonics Sonochemistry, 34*, 410–417.
28. Pagare, S., Bhatia, M., Tripathi, N., Pagare, S., & Bansal, Y. K. (2015). Secondary metabolites of plants and their role: Overview. *Current Trends in Biotechnology and Pharmacy, 9*(3), 293–304.
29. Pandey, A. (1992). Recent process developments in solid-state fermentation. *Process Biochemistry, 27*(2), 109–117.
30. Parker, C. W., Auler, A. S., Barton, M. D., Sasowsky, I. D., Senko, J. M., & Barton, H. A. (2018). Fe (III) reducing microorganisms from iron ore caves demonstrate fermentative Fe (III) reduction and promote cave formation. *Geomicrobiology Journal, 35*(4), 311–322.
31. Rul, F., Béra-Maillet, C., Champomier-Vergès, M. C., El-Mecherfi, K. E., Foligné, B., Michalski, M. C., & Savary-Auzeloux, I. (2022). Underlying evidence for the health benefits of fermented foods in humans. *Food & Function, 13*(9), 4804–4824.
32. Saisithi, P., Kasemsarn, R. O., Liston, J., & Dollar, A. M. (1966). Microbiology and chemistry of fermented fish. *Journal of Food Science, 31*(1), 105–110.
33. Şanlier, N., Gökcen, B. B., & Sezgin, A. C. (2019). Health benefits of fermented foods. *Critical Reviews in Food Science and Nutrition, 59*(3), 506–527.
34. Schäpper, D., Alam, M. N. H. Z., Szita, N., Eliasson Lantz, A., & Gernaey, K. V. (2009). Application of microbioreactors in fermentation process development: A review. *Analytical and Bioanalytical Chemistry, 395*, 679–695.
35. Schmidt, F. R. (2005). Optimization and scale up of industrial fermentation processes. *Applied Microbiology and Biotechnology, 68*, 425–435.
36. Schwachtje, J., & Baldwin, I. T. (2008). Why does herbivore attack reconfigure primary metabolism? *Plant Physiology, 146*(3), 845–851.
37. Sharma, R., Garg, P., Kumar, P., Bhatia, S. K., & Kulshrestha, S. (2020). Microbial fermentation and its role in quality improvement of fermented foods. *Fermentation, 6*(4), 106.
38. Tachie, C. Y., Onuh, J. O., & Aryee, A. N. (2024). Nutritional and potential health benefits of fermented food proteins. *Journal of the Science of Food and Agriculture, 104*(3), 1223–1233.

39. Tiwari, R., & Rana, C. S. (2015). Plant secondary metabolites: A review. *International Journal of Engineering Research and General Science, 3*(5), 661–670.
40. Vaishnav, P., & Demain, A. L. (2011). Unexpected applications of secondary metabolites. *Biotechnology Advances, 29*(2), 223–229.
41. Vandamme, E. J., & Derycke, D. G. (1983). Microbial indulines: Fermentation process, properties, and applications. In *Advances in Applied Microbiology* (Vol. 29, pp. 139–176).
42. Verma, N., & Shukla, S. (2015). Impact of various factors responsible for fluctuation in plant secondary metabolites. *Journal of Applied Research on Medicinal and Aromatic Plants, 2*(4), 105–113.
43. Verpoorte, R., Contin, A., & Memelink, J. (2002). Biotechnology for the production of plant secondary metabolites. *Phytochemistry Reviews, 1*, 13–25.
44. Vining, L. C. (1990). Functions of secondary metabolites. *Annual Review of Microbiology, 44*(1), 395–427.

CHAPTER 3

FERMENTED FOODS AND HUMAN HEALTH BENEFITS OF FERMENTED FUNCTIONAL FOODS

LEARNING OBJECTIVES

At the end of this chapter, readers will be able to:

1. Understand the role of fermentation in preserving meat and vegetable products.
2. Understand the role of fermentation in sausages and probiotic foods.
3. Explore the production techniques and fermentation in cabbage and cucumber products.
4. Understand the nutritional benefits of fermented foods in enhancing digestibility and synthesizing essential nutrients.
5. Exploring the health benefits of indigenous, lactic acid, and alkaline fermented foods and their role in preventing diseases and improving digestive health.
6. Assess the diversity and cultural significance of indigenous fermented foods globally.
7. Exploring advancements in food fermentation for enhanced nutrient production and quality.

INTRODUCTORY EXAMPLE

GT's Living Foods has built a strong reputation within the fermented foods sector, focusing on products renowned for their positive impact on well-being. Having a variety of products available, including yogurt with live cultures, kimchi packed with probiotics, and kombucha known for its digestive benefits. These products are often showcased in supermarket sections that focus on health foods, showcasing how fermentation can turn ordinary ingredients into highly nutritious options. These products offer suitable and flavorful options for consumers who prioritize digestive health and overall wellness, utilizing the beneficial effects of fermentation.

GT's Living Foods is widely recognized for its dedication to excellence and ingenuity in probiotic-rich foods, making it a standout in supermarket aisles. These products showcase the rich history of fermentation, from the smooth texture of yogurt filled with beneficial microbes to the zesty flavor of kimchi brimming with vitamins and antioxidants. Kombucha, with its bubbly charm and reputed ability to cleanse the body, complements GT's wide range of products, appealing to individuals who desire beverages that not only quench their thirst but also promote a healthy gut microbiome. With its diverse range of products, GT's Living Foods remains at the forefront of offering easily accessible options that promote health and well-being, all thanks to the skillful practice of fermentation.

UNIT INTRODUCTION

Many types of animals and plant material, which are commonly consumed as food, can also serve as a favorable environment for microbial growth. Which can undergo various fermentative processes.

The transformation of fresh raw materials using natural fermentation methods have been documented as far back as 6000 BC. Yeasts have been utilized in the production of bread and alcoholic beverages for centuries, dating back to the ancient Egyptian civilization over 5,000 years ago. The process of a fermentation has rich history, originating in the Jewish culture and then spreading to Greece and Rome. From the Middle Ages up to the present, there has been a significant improvement in the production and growth of starter productions (Rul et al., 2022).

Fermented foods have gained widespread recognition and are commonly included in culinary traditions and local eating habits. These foods are derived through fermentation using various ingredients such as fish, milk, meat, soy, tubercles rice, and corn. The production of aroma and flavor compounds encompasses a range of aldehydes, acids, ketones, alcohols, and esters. The taste of cheese results from the chemical modification of milk components by certain enzymes found in bacteria and molds. Bread, commonly made from wheat, is typically leavened through the metabolic activity of *Saccharomyces* or certain lactic acid microorganisms which can produce CO_2. Alcoholic beverages like beers, wines, and spirits are produced through fermentation process, where yeasts interact with cereals and fruits. Similarly, acetic acid bacteria such as Acetobacter and Gluconobacter plays a role in the production of vinegar (Şanlier et al., 2019).

Spices, saltwater, and microbial fermentations utilizing lactic acid bacteria (LAB) plays a vital role in protecting seafood, vegetables, and meat from harmful microbes that cause spoiling. These methods aid in acidification, which contributes to the preservation process. The interaction between salt and the acid made by the LAB has a beneficial effect on the enzymes that cause the breakdown of vegetable cells (such as sauerkraut, olives, and silage). This combination also helps to prevent oxidative changes in tissues and prevents the growth of harmful microorganisms that can spoil the food. Starter cultures can serve multiple purposes, such as transforming ingredients, acting as beneficial probiotics, or functioning as natural preservatives. It has been observed that LAB within fermented milk can generate biopeptides with potential therapeutic properties (Chaudhary et al., 2018).

Certain microorganisms can survive within the cherries of coffee, chocolate, and tea, play a crucial role in preserving their quality and imparting them with their distinctive flavors. Theobroma cacao beans undergo a process of fermentation to separate them from their pulp. Orient tempeh is made by using molds (*Rhizopus oligosporus*) on soybeans and *Monascus purpureus* on peas. In Japan, other fermented products such as soybean foods are producing through fermentation, such as soy sauce and miso (Yadav et al., 2011).

Furthermore, fermented foods can exhibit enhanced flavor profiles and textures, while also boasting extended shelf lives and heighten nutritional value when compared to their raw materials. Extensive research has been conducted on the scientific aspects of traditional fermented food production, with a focus on enhancing productivity, ensuring high quality, and maintaining safety standards, the nutritional value of fermented functional foods has been made possible through the utilization of modern scientific techniques like molecular biology, proteomics, and genetic engineering. Additionally, novel technologies like automation and microencapsulation have been played a significant role in this advancement (Shiby & Mishra, 2013). The manufacturing of fermented foods have evolved, with traditional methods being adapted for industrial-scale production. This has involved the careful selection of strains and the implementation of managed fermentation processes in modern bioreactors. These technological improvements have facilitated the production of foods made from fermentation that meet the requirements of an expanding population. This chapter will specifically address the process of fermented vegetables and meat, the health benefits associated with ingesting fermented foods, potential possibilities for functional fermented foods, and the most recent breakthroughs and trends in food fermentation (Stanton et al., 2005).

3.1. FERMENTED MEAT AND VEGETABLE PRODUCTS

Learning Objectives:

- Understand the role of fermentation in preserving meat and vegetables.
- Identify key microorganisms and processes in fermented sausage and vegetable production.

This book also covers various other topics, such as the manufacturing of nondairy probiotic drinks, milk and cheese, and commercial fermentations for alcoholic beverages. These sections provide insights into different types of fermented foods which are widely consumed by people globally. This section will provide up-to-date scientific and technological information on fermentation, focusing on important fermented foods like meat and vegetables (Stadnik & Kęska, 2015).

3.1.1. Fermented Meat

Throughout human evolution, consuming animal products has been a longstanding practice. Meat, in particular, has played a crucial role as a highly nutritious source of protein for countless generations. Meat along with its derivatives is widely consumed in diets worldwide due to its strong sensory appeal and ability to satisfy hunger. In recent years, there has been a noticeable rise in meat consumption, particularly in European countries along with China. China, in particular, stands out as the leading consumer, accounting for 25% of all animal-based food consumption worldwide.

Nevertheless, meat along with its derivatives is prone to rapid deterioration and demands meticulous storage precautions. The degradation of these substances is linked to inherent properties, including a high-water activity (Aw), a high pH, or the depletion of nutrients and minerals. These factors create optimal conditions for the proliferation of microorganisms (Lücke, 1994).

The process of meat industrialization aims to extend its shelf life, create diverse flavors, utilize less marketable components of the carcass to produce various by-products and incorporate approved condiments, spices, additives, and starter cultures. By utilizing these components, along with various physical and thermal processes, significant changes occur in the meat's physicochemical properties, resulting in enhanced sensory attributes and improved product durability (Kołożyn et al., 2009).

The preservation techniques are necessary to maintain quality of meat due to its perishability. Fermentation has proven to be an effective method of preservation. Typically, the preservation of meat involves a combination of fermentation, dehydration, and salting techniques (Papavergou et al., 2012).

3.1.1.1. Embedded Fermentation

Historical records document the long-standing use of drying, fermentation, and curing methods to preserve meat. The Mediterranean region, renowned for its favorable environment, has long been acknowledged as a center for fermenting methods. The techniques have undergone significant advancements, and today, various methods of fermentation utilize a range of microorganisms. These microorganisms play a crucial role in shaping the properties of meat products and yield a variety of unique commercialized products from fermentation (Mıynat et al., 2020).

The process of preserving fresh meat through salting and drying originated as a means of conservation when immediate consumption was not possible and alternative methods were unavailable. The production qualities of established fermentation are influenced by the geographical region, resulting in distinct flavors and textures that cater to the preferences of the local population. Additionally, these products are often recognized and celebrated for their specific regional origins.

Climatic conditions play a crucial role in the emergence of various methods for obtaining meat derivatives. In northern Europe, the practice of cooking and preserving food became more prevalent due to the colder weather conditions, which facilitated improved conservation and storage methods. The practice of dry embedment originated in southern Europe, where moderate temperatures offered greater product stability (Tran et al., 2015).

Europe, France, Italy, and Spain are the leading countries in manufacturing and using per capita for embedded fermentation. Germany is the leading consumer of fermented products, particularly sausages, with an impressive variety of 330 different types manufactured in the country. Italy has a wide variety of sausages, with approximately 1,000 different types. These products can withstand different environmental temperatures due to their fermentation processes. There has been a noticeable rise in the production and consumption of fermented products in countries like the United States, Brazil, Australia, Japan, and Great Britain (Lin et al., 2015).

3.1.1.2. Fermented Sausages

Sausages are an example of a fermented functional food. These are categorized into two major groups on an international level, depending on the manufacturing process and the final pH level of the product. Sausages from the northern regions are typically made from cow or steer meat, resulting in a more acidic flavor profile with a pH level below 5. On the other hand, sausages originating from the Mediterranean often utilize pork meat, resulting in a higher pH level above 5 and a smoother taste. A wide range of sausages is available, both commercially produced and handmade. These sausages combine minced meat, fat, salt, cure agents, sugar, and flavors. Which are then wrapped and undergo a fermentation and drying process. Sausage can be safely consumed without heating as it remains stable at environmental temperatures. Figure 3.1 provides a detailed description of the procedure for the creation of fermented sausages. It outlines the various steps involved in the production of sausages (Roca & Incze, 1990).

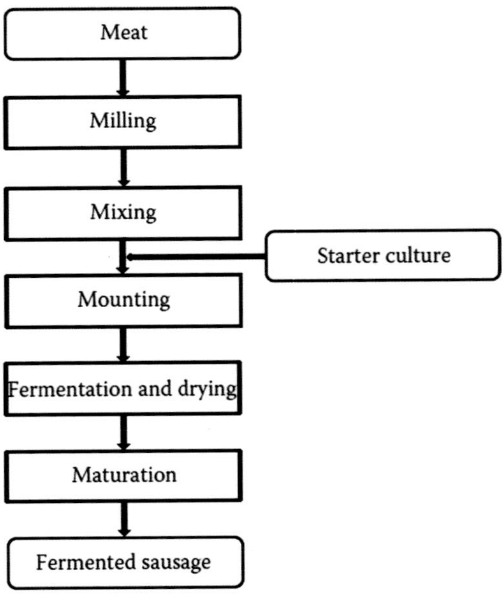

Figure 3.1. *Illustration of manufacturing flowchart of fermented sausages.*

Source: Juliano De, Creative Commons License.

Various factors contribute to the variations in the processes, including the composition of the uncooked meat material, the composition of the starter culture, the curing process, the use of mold, and the duration of drying time (Rantsiou et al., 2005). Throughout the sausage-making procedure, fermentation and drying play crucial role in enhancing texture and slice ability, preventing the growth of harmful microorganisms, creating distinct aromas and flavors, achieving the desired red color, and prolonging shelf life. All of these may be attributed to various physical, chemical, and microbiological reactions (De Vuyst et al., 2008).

The incorporation of initial culture consisting of various microorganisms imparts distinct attributes to the products. LAB plays a crucial role in lowering the pH to approximately 5, thereby inhibiting the growth of harmful bacteria. This helps in not only enhance food safety and product stability but also facilitates the drying process. Nevertheless, certain types of bacteria, including *Staphylococcus* and Kocuria, have the potential to enhance the aroma and prevent rancidity. Yeast strains (Debaryomyces) can produce enzymes that break down proteins and fats, while also metabolizing lactic acid (Holck et al., 2011). This process not only reduces the acidity of the final product but also contributes to enhancing its taste. In this particular scenario, the flavor tends to be more refined, and the scent becomes more pronounced. After the mounting step, the use of mold on a product brings about enzymatic changes to enhance the flavor, control dehydration, and limit oxygen exposure. These changes subsequently hinder the development of unwanted oxidative reactions (Comi et al., 2005). The choice of starter culture in sausage manufacturing depends on the type of cure agent used. When nitrite is utilized as the sole curing agent, it is typically advised to use LAB and *Staphylococcus*. Incorporating nitrate into the curing process necessitates the presence of nitrate-reducing bacteria of the *Micrococcus* genus within the starter culture (Figure 3.2).

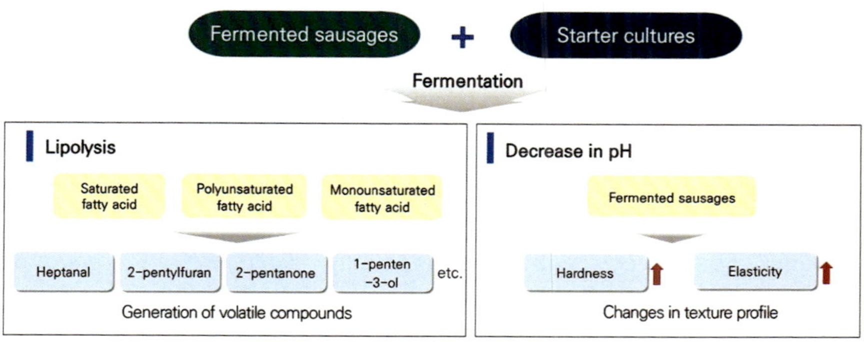

Figure 3.2. *Illustration of the effects of starter culture on quality improvement of fermented sausages.*

During sausage production, the conversion of nitrate to nitrite plays a crucial role in the development of color. This process is catalyzed by an enzyme called nitrate reductase, subsequently, the nitrite undergoes several reactions leading to the formation of nitrous oxide.

This compound then interacts with myoglobin, resulting in the formation of the intricate nitroso myoglobin complex. The intricate compound is accountable for the characteristic hue of the preserved items and helps prevent the oxidation of fats (Suzzi & Gardini, 2003).

The metabolism of Micrococcaceae releases enzymes that facilitate the breakdown of proteins and fats, resulting in enhanced texture and flavor due to the production of aromatic compounds and organic acids. *Staphylococcus* species are commonly observed as the predominant bacteria among the Micrococcaceae family (Pennacchia et al., 2004).

Starter cultures commonly consist of *S. xylosus*, *S. saprophyticus*, *S. carnosus*, and their respective subspecies carnosus and utilis. Antimicrobial compounds have garnered attention for their possible application in food production due to their ability to impede the growth of hazardous microbes such as Listeria monocytogenes and prevent food spoiling. Using a starting culture in manufacturing of meat derivatives guarantees product safety, reduces fermentation time, and improves the flavor, texture, stability, and scent, leading to a superior end product (Garriga et al., 1993).

3.1.1.3. Use of Probiotics in Meat Products

Research has shown a potential link between the consumption of sausages along the development of certain diseases, including cardiomyopathy. Efforts have been made to address this issue and enhance the positive attributes for consumer well-being. To achieve this, innovative techniques, including the incorporation of probiotic microbes, have been employed (Munekata et al., 2020).

The viability of probiotic bacteria in embedded fermentations is often compromised due to various factors such as the presence of competitor microorganisms, curing salts, and a low pH. These conditions create an unfavorable environment for probiotics. Probiotic cultures may be inhibited in the presence of certain substances such as nitrite, nitrate, and sodium chloride (Arihara, 2006).

Various strains of probiotics, including *Lactobacillus rhamnosus*, *Bifidobacterium lactis*, *Lactobacillus casei*, *Lactobacillus reuteri*, and *Lactobacillus plantarum*, may be incorporated into starter cultures to achieve the intended positive outcomes (Manassi et al., 2022).

3.1.2. Fermented Vegetables

Various methods are employed in processing of vegetables, including freezing, canning, fermenting, drying, and chemical preservation. For many people in developing countries, canned or frozen foods may be too costly or difficult to find. As a result, acid fermentation paired with salting is a common and practical method of preserving fresh vegetables. This traditional preservation technique helps in enhancing the taste and nutritional value of the vegetables and also makes them more accessible to a larger population. Bio preservation involves prolonging the period of preservation and improving the hygiene of vegetables by utilizing their inherent or introduced microorganisms and the antibacterial substances they produce. Another method of preserving vegetables is through acidification or a blend of different techniques (Breidt et al., 2007).

Did You Know?

Traditional Japanese fermented foods such as miso and natto offer not only a delightful taste to dishes but also a rich source of vital vitamins and minerals derived from the fermentation process.

Currently, peppers cabbage (sauerkraut and kimchi), and pickles (cucumber), olives, are the main vegetables that are commercially fermented. Annually, over 800 million kilograms of vegetable fermentation were produced in the US in 2006 (Jägerstad et al., 2004). On average, Germans consume about 1.8 kilograms of sauerkraut per person each year, while Syrians have a higher consumption of nearly six kilograms of fermented olives. It is worth mentioning that Koreans have a significant annual consumption of kimchi, exceeding 43 kilograms. Other vegetables such as cauliflower, carrots, artichokes, and beans are also fermented but do not hold significant economic value. In this section, one will explore the manufacturing procedures used in the production of various fermented vegetables. And also delving into the microbiological and economic aspects associated with these processes (Torres et al., 2020).

3.1.2.1. Production Techniques

The development of vegetable technologies has a lengthy historical background, spanning more than 2000 years. The vegetable fermentation process comprises a sequence of processes that are impacted by the choice of starting culture and the specific circumstances of the fermentation environment.

The production of fermented vegetables follows a similar process, where sugars are transformed into acids, resulting in the development of distinct characteristics in the final product. This manufacturing technique shares common principles with other lactic acid fermentations. Unlike the dairy business, the fermented vegetable sector still uses nonstarter lactic microorganisms (autochthonous) for the process of fermentation. To facilitate lactic acid fermentation, it is important to create the appropriate environment by considering factors such as salt and oxygen concentrations, as well as temperature. This is crucial due to the significant difference in microflora found in raw vegetables, particularly the abundance of lactic acid bacteria compared to other types (Adewale et al., 2015).

The prior treatment step is a crucial stage that encompasses various operations to carefully select and clean the raw vegetables. In industrial settings, specific treatments like blanching, peeling, and/or cutting are frequently employed (Ching et al., 2017).

Creating an optimal environment near the vegetable is crucial to ensure the growth and dominance of beneficial microorganisms. In many types of vegetable fermentations, the addition of salt is very crucial. It helps in preventing the growth of spoilage microorganisms, aids in the removal of excess water from the vegetables, and adds a salty and flavorful touch to the final product. The variation in salt concentrations utilized during production contributes to the diversity of LAB biotypes that thrive in different fermentation environments. Using starter cultures is a technique that helps enhance the growth of beneficial microorganisms in the fermentation process of vegetables. The LAB utilized for this objective consisting various species such as *Lactobacillus*, *Leuconostoc mesenteroides*, and *Lactococcus*. The starter cultures can quickly colonize the substrate, thrive in different environmental conditions, and initiate the fermentation process of the vegetable (Fricke & Emmerling, 1998).

Several key factors play a significant role in the fermentation process of vegetable, including temperature, pH levels, and the presence of low oxygen concentrations. The process typically involves using temperatures between 10°C and 38°C (Hainsworth & Uhure, 2007). The temperature required for the fermentation process is determined by the prevailing microorganisms involved. The buffer systems facilitate the use of sugars through the starter culture. Ensuring optimal pH levels and neutralizing acidity during fermentation has been a common practice to ensure the full utilization of sugars. To maintain the longevity of fermented vegetables, it is crucial to ensure that the pH level remains below 4. To preserve anaerobic conditions, the vegetable must be completely submerged in the brine within the vessels. Industrial anaerobic tanks offer a more favorable environment for anaerobic conditions compared to other vessels (Rigby & Essletzbichler, 2006).

3.1.2.2. Cabbage Fermentation

Sauerkraut, also known as kraut, is a product that is formed through the process of lactic acid fermentation. It is made by shredding fresh white cabbage and then salting it. In the past, sauerkraut played a crucial role in the diets of naval forces and seafarers. Due to limited access to fresh produce, which would have provided vitamin C to avoid scurvy, sauerkraut became a valuable source of nutrition (Tolonen et al., 2002).

The manufacturing processes of sauerkraut are illustrated in the flowchart provided in Figure 3.3. The cabbage that arrives at the factory is taken into a coring-scoring machine, while the cored heads are then moved to a turning table in which they are carefully trimmed, and any damaged areas are removed. The cabbage is shredded and then carried to the fermentation vat.

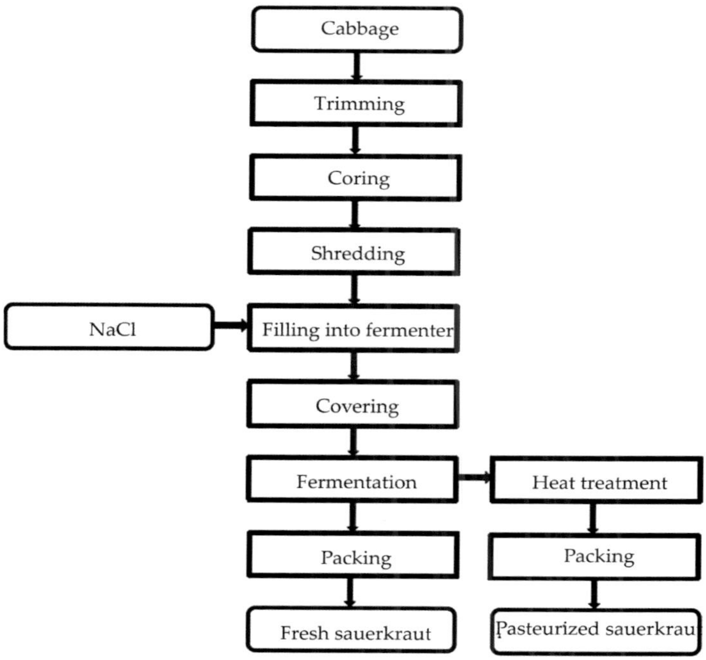

Figure 3.3. *Illustration of manufacturing flowchart of fresh and pasteurized sauerkraut.*

Source: Ana Lúcia, Creative Commons License.

A small amount of salt, around 2 to 3%, is carefully sprinkled over the shreds and spread out in the vat. The juice is extracted from the cabbage shortly after the salt is added. Once the vat is full, a sturdy lid is placed on top, and the fermentation process begins, lasting for a minimum of one month at a temperature below 15.5°C. The fermentation process is considered finished once the shreds have been properly treated and the titratable acidity, measured in terms of lactic acid, reaches the level of 1.5%. After completing the necessary steps, the kraut and brine are carefully transferred into retail packages. The product might be canned, thermally processed at around 75°C (pasteurized sauerkraut), or packed for non-pasteurized, refrigerated marketing in plastic bags that are sealed (polybags) with or without approved preservative ingredients (Jagannath et al., 2012).

The process of fermenting sauerkraut is a fascinating microbiological phenomenon that can be divided into two distinct stages. The first stage involves a diverse range of microorganisms, while the second stage is dominated by the microorganisms naturally present in the raw cabbage. The fermentation process begins with conversion of sugar into lactic acid, acetic acid, alcohol, and CO_2 by *L. mesenteroides*. Carbon dioxide plays a crucial role in creating the ideal anaerobic environment required for the fermentation of cabbage. Following this, the growth of *L. mesenteroides* is hindered by the buildup of CO_2 in the vat, and the fermentation process proceeds using *Lactobacillus brevis*, *Pediococcus cerevisiae*, and ultimately *Lb. plantarum* (Cho & Park, 2006).

Kimchi is passed down through generations in Korean culture, by fermenting cabbage and vegetables. Kimchi is popular in the Far East as a widely consumed fermented vegetable. The production of kimchi involves using fresh cabbage which is divided in half or shredded. It is then soaked in a brine with a salt concentration of around 10% overnight, before being washed and drained. The kimchi is carefully sealed in a jar and gently pressed to ensuring all ingredients are fully submerged in the flavorful juice. *L. mesenteroides* serves as the prevailing microorganism during the initial stages of ripening. The dominance of *Lactobacillus* species in the final phases of kimchi fermentation is contingent upon the temperature conditions. The bacterial communities in different types of kimchi vary in diversity and richness, which can be attributed to variations in ingredients and manufacturing processes (Martinez et al., 2012).

One notable distinction between sauerkraut and kimchi is the fact the latter incorporates additional ingredients beyond cabbage and salt. Kimchi is commonly prepared using cabbage, radishes, and cucumbers. Various spices and flavors are often incorporated into kimchi, according to the specific variety (Weaver et al., 1992).

3.1.2.3. Cucumber Fermentation

Process of combining acid, spices, and sugar with cucumbers (Cucumis sativus L.) results in the creation of pickles, a popular acidic food product. Pickled cucumbers are commonly produced in various regions across Africa, Asia, and Latin America. There are many ways to categorize pickled products, depending on the ingredients and preparation methods employed: There are two types of pickles: fermented pickles and fresh-pack pickles. Fermented pickles are made through a natural curing process, while fresh-pack pickles are made by adding vinegar to achieve a pH of 4.6 or below (Fleming, 1984).

For the production of fermented pickles, carefully chosen cucumbers are immersed in a brine solution containing around 5% salt. The brine power is incrementally raised to 16% through this process. Methods have been devised for pickling cucumbers in sealed oxygen-free containers using significantly reduced amounts of salt. With this technique, it becomes possible to ferment and store cucumbers at lower salt concentrations without desalting (Franco et al., 2012).

Several microorganisms play a crucial role in the pickling process, typically from around day 7 to day 14. At a pH of 3.2, the metabolic activity of *Lb. plantarum* remains hindered, resulting in the completion of fermentation.

The preservation of the product can be achieved through various methods such as fermentation, pasteurization, or refrigeration. Additionally, it may include sweeteners, spices, flavorings, and other approved ingredients (Etchells et al., 1964).

3.2. HUMAN HEALTH BENEFITS OF FERMENTED FOODS

Learning Objectives:

- Understand the nutritional and health benefits of consuming fermented foods.
- Understand the benefits of indigenous, lactic acid and alkaline fermented foods.

Numerous food options preferred by individuals are abundant in essential nutrients and have a limited shelf life in their unprocessed state. Typically, these foods have a significant amount of moisture and water activity, which creates an environment conducive to the growth of microorganisms. Microbiological deterioration is the significant issue due to various factors, and people have discovered ways to prevent food deterioration and capitalize on foods have been altered through fermentation. This results in the production of nutritious and delicious products (Farnworth, 2005). Over the years, there has been a gradual realization of the role played by microscopic organisms in the spoiling and fermentation of food and beverages. Interestingly, these organisms have been a part of our daily diet for a long time (Marco et al., 2017). Microorganisms in food items often evoke concern and anxiety, as people contemplate the presence of potentially harmful bacteria, toxins, and other alarming issues frequently highlighted in the media. It is surprising to discover number of individuals are unaware of the advantages of microorganisms in food and are often taken aback to learn that everyday items like cheese, salami, pickles, yogurt, and various other food items contain living organisms that are consumed as part of their regular diet (Voidarou et al., 2020).

Microorganisms in foods can break down complex food compounds, such as polysaccharides, proteins, oligosaccharides, and other minor components. In certain cases, these components can be toxic or difficult to digest for humans. Various types of seeds, like pulses, can pose this particular issue, while cassava roots being toxic if not processed correctly, which involves the help of microorganisms. Various microorganisms, such as bacteria and fungi, have significant impact on the production of fermented foods. These foods, like moldy cheeses, salami, yogurts, pickles, and fermented cassava flours, are widely consumed and can be a valuable source of nutrition. The enzymatic activity of these tiny organisms within vegetable and animal-based foods typically enhances their digestibility. Microorganisms play a crucial role to enhance nutritional quality of vegetable-based foods by synthesizing vitamins, other beneficial compounds, and high-quality proteins (Asghar et al., 2017).

Foods and drinks made from fermentation have been consumed by humans for centuries and continue to be a part of our diets today. Traditional Oriental dishes incorporate fermented foods and ingredients sourced from soybeans, fish, and other raw materials, similar to how Western countries utilize pickled vegetables, yogurts, cheeses, and more. Alcoholic fermented drinks and dairy foods that are fermented are widely recognized as significant examples of fermented goods that are both well-received and have potential health benefits. While the use of ethanol is often seen as risky, the presence of other components in wine has been found to play a crucial role in promoting a healthy lifestyle. Artisanal beers are highly valued in various African countries due to their rich content of B-complex vitamins. It is important to take the extended shelf life that fermentation provides for highly perishable foods. These types of nearly fresh foods contain partially consumed components that are highly valuable in human diets (Mukherjee et al., 2024).

While numerous benefits have been discussed regarding the nutritional value of fermented foods, while certain challenges exist beyond the inclusion of ethanol in fermented alcoholic beverages. One limitation are related to this kind of processed food is the elevated sodium content found in certain pickled vegetables and fermented meat products (Rai & Jeyaram, 2015).

3.2.1. Indigenous Fermented Foods

Fermentation has been used for centuries as a cost-effective method of producing and preserving various types of food. The raw material undergoes microbial and enzymatic transformations, resulting in changed nutritional value, texture, flavor, and appearance. The food is deemed spoiled if the final product is unpleasant in taste, has an off-flavor, or is potentially harmful. However, if not, this is referred to as a food that has undergone fermentation (Ashenafi, 2006).

Steinkraus conducted a comprehensive study on fermented foods from various countries, categorizing them based on the types of fermentation and raw materials used:

- Protein-rich vegetarian meat substitutes like Indonesian tempeh;
- Fermented foods that involve lactic acid fermentation, such as acid-fermented fruits and vegetables and acid-fermented milk;
- Alkaline-fermented foods like Japanese natto;
- Fermented foods where ethanol is a major product, like vinegar, alcoholic beverages, and acetic acid fermentations;
- Various types of sauces and pastes have meat-like flavors, which include Chinese soy sauce and Japanese shoyu and miso;
- Mushrooms made on agricultural lignocellulosic wastes are a source of microbial protein. The widely recognized categorization of fermented foods is being embraced globally. (Ramos & Schwan, 2017).

From a scientific and technological perspective, fermentation serves at least five important functions:

- It enhances the flavors and aromas of foods, resulting in a more enjoyable culinary experience, alters the texture of food, and helps produce a more diverse and nutritious diet;
- Another important aspect of fermentation is its role in food preservation. Various types of fermentations, including lactic acid, alcoholic, acetic acid, and alkaline fermentations, help to extend the shelf life of food products;
- It enhances food substrates by adding protein, essential amino acids, essential fatty acids, and vitamins;
- It detoxifies specific raw materials to make them safe for consumption;
- It reduces cooking times and fuel needs (Egwim et al., 2013).

Artificial vitamins, and other nutrients are commonly added to foods in many western countries. However, in the developing world, the use of fermentative processes for bio-enrichment is crucial, especially for vitamins and essential amino acids.

Steinkraus examined the global industrialization of various traditional fermented foods. These included Japanese shoyu and miso, which are soybean sauces and pastes, as well as sake and natto. The study also covered Malaysian tapai, made from glutinous rice or cassava roots, and Africa's Indigenous-brewed beersmade from sorghum, millet, or corn. Additionally, gari, a fermented food produced from cassava roots, was included in the analysis. Various factors are evaluated when assessing the achievements of certain fermented foods produced on an industrial scale. This includes popular items like yogurts, soy sauces, miso, brewing, winemaking, cheeses, and vinegar. On the other hand, there are also prospects for expanding village-level techniques to larger-scale industrial production with products like tempeh and indigenous fermented beverages (Ramos et al., 2023).

According to Blandino, fermented foods are made across the world using various microorganisms and raw materials. The fermentation processes can be categorized into four primary types: Various types of fermentation includes processes that involve the production of alcohol, lactic acid, acetic acid, and alkaline substances. The authors conducted a comprehensive review of cereal-based fermented foods and beverages, recognizing the significant role of cereal grains in providing essential nutrients in diets across the globe. The nutritional value, sensory properties, and functional qualities of these primary products can be enhanced by fermentation processes. All while keeping costs low. This is because fermentation helps address deficiencies in amino acids and vitamins (Aidoo et al., 2006).

Legumes play a crucial role in Egypt and other developing countries as valuable protein sources, especially in light of the expensive nature of meat. The fermentation processes linked to legume grains enhance digestibility, nutritional value, and flavor,while ensuring food safety. Indonesian tempeh serves a fascinating fermented food with a mild taste, abundant in protein, and widely embraced by vegetarians in Western countries like the Netherlands and the US. In their research, Nassar examined the production of tempeh on a small scale in Egypt and utilized a combination of legumes that were readily accessible in the local area, including green pea, and lupine broad bean, chickpea. The authors were able tocreate tempeh using different raw materials, resulting in a higher nutritional value and reduced levels of oligosaccharides successfully that can cause flatulence (Narzary et al., 2016).

3.2.2. Lactic Acid-Fermented Foods

Throughout history, fermentation has been a crucial method for preserving animal and vegetable-based items, ensuring the availability of high-quality and nutritious foods. These time-honored recipes have been passed down through generations in China. The authors emphasize the benefits of fermentation, including enhanced safety, increased nutritional value, and improved flavors of foods. It is must to acknowledge the significance of fermented goods in developing nations with large populations, which play a vital role in providing safe food and enhancing its sensory and nutritional qualities. LABs play a significant role in various food fermentations are widely recognized as safe and are also known for their association with the probiotic properties found in certain commercial products (Olasupo et al., 2002).

The use of LAB as a fermentation agent in various food products has been widely documented. Liu has conducted an extensive analysis of four primary types of fermentation based on commonly consumed ingredients in Chinese diets: dairy, vegetables, soybeans, and meat. The authors highlight the importance of enhancing flavor and the release of chemically active substances using the metabolism process of microorganisms, which can provide additional health benefits for humans.

According to Steinkraus, fermentations that produce lactic acid are typically considered safe and rely a transformation of carbohydrates towards lactic acid by various types of LAB. One important factor that offers dietary advantages to consumers, particularly in developing countries, is the longer shelf life of consumable raw materials due to lactic acidification (Bousquet et al., 2021).

In Brazil, the expanded salty biscuits made from fermented and sundried cassava starch or sour cassava starch have a distinct flavor that is highly regarded by consumers. This traditional fermented cassava starch is commonly made in Brazil and, to a lesser extent, in some other South American nations. It is referred to as polvilho azedo in Brazil almidón agrio in Colombia. The process for natural fermentation of lactic acid takes place throughout around 20 to 40 days resulting in the production of organic acids that subsequently lower the pH. Following this timeframe, the acidic starch is left to dry in the sun for approximately 1 to 2 days. This starch is highly valued for its technological potential in creating delicious expanded baked goods, such as salty biscuits and cheese bread, without the need for extrusion. These foods can easily be made in the comfort of your own home, and the food items are free from gluten (Jespersen, 2003) (Figure 3.4).

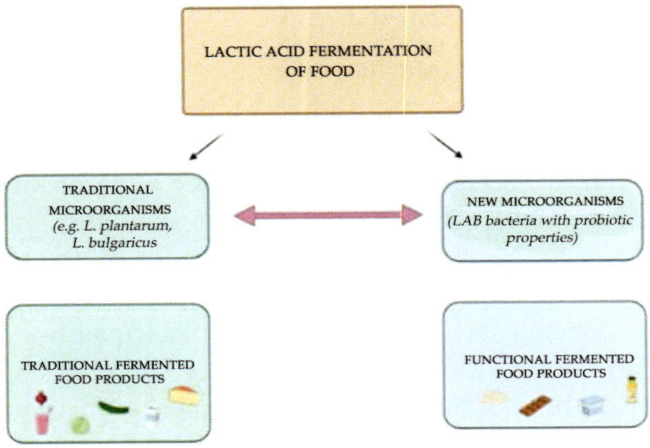

Figure 3.4. *Illustration of lactic acid microorganisms used in fermented foods.*

Source: Barbara Sonek, Creative Commons License.

3.2.3. Alkaline-Fermented Foods

Alkaline-fermented foods are a lesser-known group of food products that gained popularity in Southeast Asia and African countries. One example of a widely recognized product is Japanese natto. It is produced by fermenting dehulled and partially prepared soybean seeds with Bacillus spp., primarily Bacillus subtilis. This fermentation process leads to an increase pH levels, ranging from 8 to 9, due to the release of ammonia. As a result, the product has a distinct and potent smell, reminiscent of ammonia (Ouoba et al., 2010).

Alkaline fermentation is also known to enhance digestibility by breaking down proteins into amino acids and peptides. Additionally, there have been findings of increased synthesis of vitamins such as riboflavin and vitamin B12 during this process. Notably, natto has been found to contain three to five times higher concentrations of these vitamins compared to cooked soybeans. This process of alkaline fermentation of African locust legumes is known to have a significant impact on detoxification. These beans, which are initially poisonous to humans, become edible after undergoing this fermentation process. The resulting food, known as dawadawa, is safe for consumption (Parkouda et al., 2009).

3.3. FERMENTED FUNCTIONAL FOODS

Learning Objectives:

- Understand the concept of fermented functional foods and their potential health benefits.
- Explore the role of probiotics in fermented foods and their mechanisms of action in promoting health.

Fermented foods have been produced for centuries and are considered one of the earliest food-processing methods in human history. One growing body of research is suggesting fermented foods have the potential to provide various health benefits. Typically, functional foods are acknowledged for their ability to enhance overall health and lower the risk of certain diseases. In 1984, the concept of "functional food" emerged as the Japanese started exploring the connections between nutrition, sensory satisfaction, fortification, and the impact on physiological systems. According to Roberfroid, functional foods are those that fulfill basic nutritional needs and also have the potential to offer additional health benefits (Stanton et al., 2005). These foods, although look like regular food, have been modified to serve specific physiological purposes when ingested as part of a regular diet.

3.3.1. Classes of Functional Foods

According to Spence (Gul et al., 2016), four primary kinds of functional foods available in the market. Several types of products fall under the category of fermented functional foods. These include fortified products, enriched products that utilize fermented altered products, probiotics, and prebiotics that replace existing components with beneficial ones, and enhanced commodities that undergo changes in nutrient composition, such as high-lysine corn and golden rice).

3.3.2. Probiotics as Fermented Functional Foods

In recent years, the industrial segment of probiotic-fermented foods has experienced significant growth. It is worth to note g that probiotic products have the potential to be incorporated into daily routines as a form of preventive medicine for various age groups (Syngai et al., 2016).

Probiotics are described as live microorganisms by FAO that provide therapeutic or health benefits when consumed in sufficient quantities. The probiotic field primarily focuses on LAB and bifidobacteria uses, while less common genera include Aspergillus, Bacillus, and *Saccharomyces* (Beena et al., 2012).

Various dairy products such as fermented milk, yogurts, buttermilk, and cheese, as well as fermented meats, vegetables, soft drinks, and cereals, have utilized the beneficial properties of *Lactobacillus* and *Bifidobacterium* in their production. Several microorganisms, including *Bacillus, Enterococcus, Aspergillus*, and *Saccharomyces*, have utilized in feed-animal preparations in previous studies (Kumar et al., 2022).

Practice Problem

Discuss the role of probiotics in fermented foods. How do probiotics contribute to human health, and what are their mechanisms of action?

Solutions to Practice Problem

Probiotics are beneficial microorganisms found in fermented foods. They contribute to human health by promoting a balanced gut microbiota, enhancing digestion and nutrient absorption, supporting immune function, and potentially reducing inflammation. Mechanisms of action include competitive exclusion of pathogenic bacteria, production of antimicrobial compounds, and modulation of immune responses in the gut mucosa.

Probiotic products are often touted for their potential to prevent inflammatory diseases, allergies, increased cholesterol levels, digestive problems, and certain types of cancer. The consumption of LAB probiotic functional foods that included *Bifidobacterium longum*, *B. lactis*, *Lactobacillus acidophilus*, and *L. rhamnosus* has been extensively studied, and one of the suggested benefits is the modulation of the host immune system. Probiotics are beneficial for individuals with food allergies, lactose intolerance, or inflammatory conditions like inflammatory bowel diseases (IBD). Which can also help prevent infections caused by pathogens like *Salmonella*, *Escherichia coli*, and *Helicobacter pylori* by competing for receptor sites and producing bacteriocins. *Saccharomyces boulardii* provides a protective effect against toxin A, produced by *Clostridium difficile*, which can potentially lead to intestinal damage and inflammation (Mills et al., 2009).

Multiple studies have indicated that LAB probiotics may have a beneficial effect in preventing cancer by promoting various immune responses and reducing factors. Probiotics have also been associated with benefits such as allergy protection and defense against the effects of mycotoxins (Siciliano et al., 2021).

3.3.3. Health Benefits and Mode of Action of Probiotics Produced by Fermented Foods

Numerous health advantages have been linked to probiotics, including immunity stimulation, improved digestion, protection against infection, potential antitumor properties, relief from atopic dermatitis and allergy symptoms, decreased cholesterol, and favorable effects on constipation and certain bowel diseases. Several potential mechanisms of action have been proposed: (a) Interactions with other kinds of microbes and the balancing of gut bacteria; (b) Exploring the potential of utilizing fermented functional foods in the treatment and prevention of diarrhea by targeting the receptors of pathogenic microbes and rotavirus; (c) Suppressing the growth of harmful bacteria by producing antimicrobial chemicals, H_2O_2, acids, and bacteriocins; (d) Altering the pH of the intestines through the production of lactic acid and short-chain fatty acids; (e) Engaging in a constant battle for essential nutrients and generating beneficial growth factors within the gastrointestinal tract; (f) Stimulating immunomodulation through the induction of lymphocytes, immunoglobulins IgA, IgM, and IgE, the anti-inflammatory cytokine IL-10, and interferon IFN-γ; (g) Decreasing the mutagenic activity and procarcinogenic elements in the colon, as well as reducing the levels of fecal enzyme activities such as nitratoreductase, azoreductase, and glucuronidase; (h) Triggering metabolic effects through the production of lactase and other hydrolytic enzymes, including proteases and lipases; (i) Investigating the potential benefits of specific LAB strains in mitigating *H. pylori* infection, reducing ulcer-inducing factors, and providing protection against induced gastritis; and (j) Lowering cholesterol levels. It should be noted, all of the mentioned advantages can be attributed to every probiotic strain, as certain positive responses are specific to certain strains. There is substantial body of evidence has been documented, with a majority of the findings focused on individuals with existing health conditions. However, it is important to establish the health advantages in the general population, specifically those who are generally healthy (Jalili et al., 2023).

Did You Know?

Probiotics present in fermented meals can endure the digestive tract, which could potentially provide health advantages by establishing a colony of good bacteria in the gut.

Recent research has indicated that probiotics can transform milk protein into bioactive peptides, potentially leads to a beneficial impact on hypertension. The observed effect is linked in inhibiting certain enzymes, including dipeptidyl carboxypeptidase, also known as angiotensin, I-converting enzyme (ACE). The enzymes facilitate the transformation of angiotensin I into the powerful vasoconstrictor angiotensin II, triggering the release of aldosterone and an elevation in sodium levels, resulting in a rise in blood pressure. Studies have been conducted on living organisms have shown that ACE inhibitor peptides can effectively lower blood pressure when given through intravenous or oral administration (Voidarou et al., 2020) (Figure 3.5).

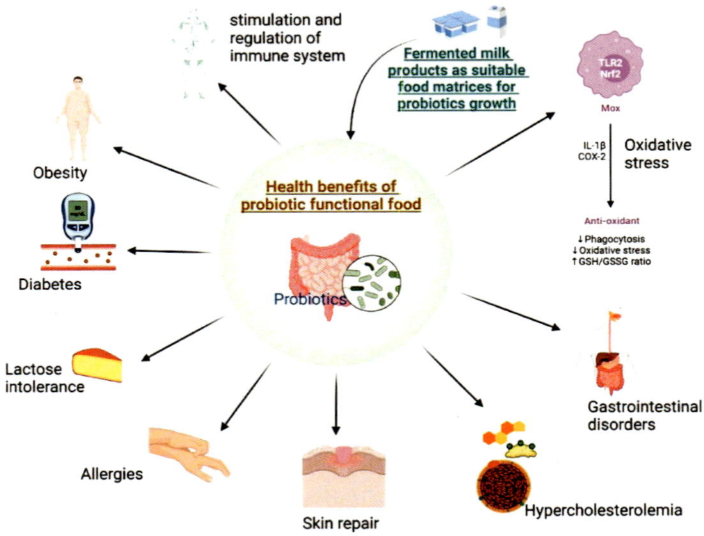

Figure 3.5. *Illustration of the health benefits of the probiotics food.*

Source: Machado Prado, Creative Commons License.

3.3.4. Food Matrices as Vehicles for Probiotics

Various food products such as cheese, butter, cereal, mayonnaise, meat, powdered milk for infants, and vegetables can be used as probiotics carriers. Dairy products are widely recognized for their ability to provide beneficial bacteria to the gastrointestinal tract. This is because dairy products contain functional substances like milk proteins that both protect and interact with the probiotics, acting as a buffer for the bacteria in the stomach. In addition, ice cream and frozen dairy-based desserts can serve as effectiveprobiotics carriers. Their ability to be kept at low temperatures and maintain good viability upon consumption makes them convenient options for people of all ages (Flach et al., 2018).

Some examples of nondairy matrices include fruits, legumes, vegetables, and cereals. These matrices are known for their low-fat content and lack of lactose. A significant portion of these products consists of non-dairy fermented beverages made from vegetables and fruits. Nevertheless, fruit juices have shown certain limitations due to the need for protecting probiotics from acidic conditions. One way to address this limitation is by utilizing microencapsulation or nanoencapsulation technologies. These advanced techniques create a protective coating that shields probiotic bacteria from the external environment, effectively overcoming the drawbacks. Commonly used food-grade polymers for microencapsulation include alginate, chitosan, pectin carboxymethyl cellulose (CMC), and carrageenan gelatin (Cunningham et al., 2021).

There are various methods for incorporating probiotic microorganisms into the production of fermented products. One method is to add probiotic microorganisms along with the starter cultures directly into the vat. Another approach is introducing the probiotic microorganisms in milk separately from the starter cultures. Alternatively, the probiotic microorganism may be included as a starter culture once the fermentation stages have finished. In the last scenario, the duration of fermentation is typically longer compared to conventional methods that utilize non probiotic starter cultures. The regulations governing functional foods require probiotics maintain their viability throughout the production process and remain stable cum viable during storage and use, and effectively survive in the intestinal ecosystem. Additionally, the host must be able to derive the benefits of consuming probiotics (Sanders & Marco, 2010).

SUMMARY

- The chapter on "Fermented Foods and Health Benefits" provides comprehensive exploration on fermented food varieties and their impacts on human health.
- It covers fermented meat and vegetable products, detailing their production processes and cultural significance. The discussion extends to indigenous fermented foods and alkaline-fermented varieties, highlighting their unique properties and regional variations.
- Emphasis is placed on fermented foods as functional foods, elucidating their role in promoting health beyond basic nutrition through probiotics and other bioactive compounds.

REVIEW QUESTIONS

1. Describe fermentation in meat products and identify key microbes.
2. Compare the health benefits of fermented vegetables vs. dairy products.
3. Discuss the cultural significance of indigenous fermented foods.
4. Explain functional fermented foods and their benefits.
5. How do probiotics in fermented foods benefit gut health and immunity?
6. Explore technological advances in fermented food production.
7. Challenges in using fermented foods for probiotics and solutions.
8. Production and health benefits of bioactive peptides in fermented milk.
9. Role of vitamins in fermented foods and their enhancement.
10. Summarize the health benefits and integration of fermented foods.

MULTIPLE CHOICE QUESTIONS

1. What are the main health benefits associated with fermented foods?

a. Increased cholesterol levels and weight gain
b. Improved digestion and enhanced nutrient absorption
c. Reduced vitamin content and weaker immune response
d. Allergic reactions and gastrointestinal discomfort

2. Which category of fermented foods focuses on their role in promoting health beyond basic nutrition?

a. Indigenous fermented foods

b. Functional foods

c. Alkaline-fermented foods

d. Fermented meat products

3. What role do probiotics play in fermented functional foods?

a. They increase sugar content

b. They enhance food color

c. They improve shelf life

d. They support gut health and immune function

4. Which technological advancement is highlighted in the chapter regarding food fermentation?

a. Increased manual labor in production

b. Advanced machinery for faster fermentation

c. Use of traditional fermentation methods

d. Improved stability and efficacy of probiotics

5. What is a characteristic feature of alkaline-fermented foods discussed in the chapter?

a. High acidity

b. Low salt content

c. Bitter taste

d. Elevated pH levels

Answers to Multiple Questions

1. (b); 2. (b); 3. (d); 4. (d); 5. (d)

REFERENCES

1. Adewale, P., Dumont, M. J., & Ngadi, M. (2015). Recent trends of biodiesel production from animal fat wastes and associated production techniques. *Renewable and Sustainable Energy Reviews, 45*(5), 574–588.

2. Aidoo, K. E., Nout, M. J., & Sarkar, P. K. (2006). Occurrence and function of yeasts in Asian indigenous fermented foods. *FEMS Yeast Research, 6*(1), 30–39.

3. Arihara, K. (2006). Strategies for designing novel functional meat products. *Meat Science, 74*(1), 219–229.

4. Asghar, F., Ali, S., Goraya, A., Javaid, I., & Hussain, Z. (2017). A review on the role of fermented foods as health promoters. *International Journal of Scientific Research in Science and Technology, 3*(1), 141–148.

5. Ashenafi, M. (2006). A review on the microbiology of indigenous fermented foods and beverages of Ethiopia. *Ethiopian Journal of Biological Sciences, 5*(2), 189–245.

6. Beena Divya, J., Kulangara Varsha, K., Madhavan Nampoothiri, K., Ismail, B., & Pandey, A. (2012). Probiotic fermented foods for health benefits. *Engineering in Life Sciences, 12*(4), 377–390.

7. Bousquet, J., Anto, J. M., Czarlewski, W., Haahtela, T., Fonseca, S. C., Iaccarino, G., & Constantinidis, J. (2021). Cabbage and fermented vegetables: From death rate heterogeneity in countries to candidates for mitigation strategies of severe COVID-19. *Allergy, 76*(3), 735–750.

8. Breidt, F., McFeeters, R. F., & Díaz-Muñiz, I. (2007). Fermented vegetables. In *Food Microbiology: Fundamentals and Frontiers* (pp. 783–793).

9. Chaudhary, A., Sharma, D. K., & Arora, A. (2018). Prospects of Indian traditional fermented food as functional foods. *The Indian Journal of Agricultural Sciences, 88*(10), 1496–1501.

10. Ching, S. H., Bansal, N., & Bhandari, B. (2017). Alginate gel particles—A review of production techniques and physical properties. *Critical Reviews in Food Science and Nutrition, 57*(6), 1133–1152.

11. Cho, H. S., & Park, B. H. (2006). Physicochemical characteristics of cabbage Kimchi during fermentation. *Korean Journal of Food and Cookery Science, 22*(5), 600–608.

12. Comi, G., Urso, R., Iacumin, L., Rantsiou, K., Cattaneo, P., Cantoni, C., & Cocolin, L. (2005). Characterization of naturally fermented sausages produced in the North East of Italy. *Meat Science, 69*(3), 381–392.

13. Cunningham, M., Vinderola, G., Charalampopoulos, D., Lebeer, S., Sanders, M. E., & Grimaldi, R. (2021). Applying probiotics and prebiotics in new delivery formats—Is the clinical evidence transferable? *Trends in Food Science & Technology, 112*, 495–506.

14. De Vuyst, L., Falony, G., & Leroy, F. (2008). Probiotics in fermented sausages. *Meat Science, 80*(1), 75–78.

15. Dziuba, B., & Dziuba, M. (2014). Milk proteins-derived bioactive peptides in dairy products: Molecular, biological and methodological aspects. *Acta Scientiarum Polonorum Technologia Alimentaria, 13*(1), 5–26.

16. Egwim Evans, A. M., Abubakar, Y., & Mainuna, B. (2013). Nigerian indigenous fermented foods: Processes and prospects. In *Mycotoxin and Food Safety in Developing Countries* (pp. 153–180).
17. Etchells, J. L., Costilow, R. N., Anderson, T. E., & Bell, T. A. (1964). Pure culture fermentation of brined cucumbers. *Applied Microbiology, 12*(6), 523–535.
18. Farnworth, E. R. (2005). The beneficial health effects of fermented foods—Potential probiotics around the world. *Journal of Nutraceuticals, Functional & Medical Foods, 4*(3–4), 93–117.
19. Flach, J., van der Waal, M. B., van den Nieuwboer, M., Claassen, E., & Larsen, O. F. (2018). The underexposed role of food matrices in probiotic products: Reviewing the relationship between carrier matrices and product parameters. *Critical Reviews in Food Science and Nutrition, 58*(15), 2570–2584.
20. Fleming, H. P. (1984). Developments in cucumber fermentation. *Journal of Chemical Technology and Biotechnology, 34*(4), 241–252.
21. Franco, W., Pérez-Díaz, I. M., Johanningsmeier, S. D., & McFeeters, R. F. (2012). Characteristics of spoilage-associated secondary cucumber fermentation. *Applied and Environmental Microbiology, 78*(4), 1273–1284.
22. Fricke, J., & Emmerling, A. (1998). Aerogels—Recent progress in production techniques and novel applications. *Journal of Sol-Gel Science and Technology, 13*, 299–303.
23. Garriga, M., Hugas, M., Aymerich, T., & Monfort, J. M. (1993). Bacteriocinogenic activity of lactobacilli from fermented sausages. *Journal of Applied Bacteriology, 75*(2), 142–148.
24. Guha, S., Sharma, H., Deshwal, G. K., & Rao, P. S. (2021). A comprehensive review on bioactive peptides derived from milk and milk products of minor dairy species. *Food Production, Processing and Nutrition, 3*, 1–21.
25. Gul, K., Singh, A. K., & Jabeen, R. (2016). Nutraceuticals and functional foods: The foods for the future world. *Critical Reviews in Food Science and Nutrition, 56*(16), 2617–2627.
26. Hainsworth, S. V., & Uhure, N. J. (2007). Diamond like carbon coatings for tribology: Production techniques, characterization methods and applications. *International Materials Reviews, 52*(3), 153–174.
27. Herbers, K. (2003). Vitamin production in transgenic plants. *Journal of Plant Physiology, 160*(7), 821–829.
28. Holck, A. L., Axelsson, L., Rode, T. M., Høy, M., Måge, I., Alvseike, O., &

Heir, E. (2011). Reduction of verotoxigenic Escherichia coli in production of fermented sausages. *Meat Science, 89*(3), 286–295.

29. Jagannath, A., Raju, P. S., & Bawa, A. S. (2012). A two-step controlled lactic fermentation of cabbage for improved chemical and microbiological qualities. *Journal of Food Quality, 35*(1), 13–20.
30. Jägerstad, M., Jastrebova, J., & Svensson, U. (2004). Folates in fermented vegetables—A pilot study. *LWT–Food Science and Technology, 37*(6), 603–611.
31. Jalili, M., Nazari, M., & Magkos, F. (2023). Fermented foods in the management of obesity: Mechanisms of action and future challenges. *International Journal of Molecular Sciences, 24*(3), 2665.
32. Jespersen, L. (2003). Occurrence and taxonomic characteristics of strains of *Saccharomyces cerevisiae* predominant in African indigenous fermented foods and beverages. *FEMS Yeast Research, 3*(2), 191–200.
33. Kołożyn-Krajewska, D., & Dolatowski, Z. J. (2009). Probiotics in fermented meat products. *Acta Scientiarum Polonorum Technologia Alimentaria, 8*(2), 61–74.
34. Korhonen, H., & Pihlanto, A. (2006). Bioactive peptides: Production and functionality. *International Dairy Journal, 16*(9), 945–960.
35. Kumar, D., Lal, M. K., Dutt, S., Raigond, P., Changan, S. S., Tiwari, R. K., & Singh, B. (2022). Functional fermented probiotics, prebiotics, and symbiotic from non-dairy products: A perspective from nutraceuticals. *Molecular Nutrition & Food Research, 66*(14), 2101059.
36. Latorre-Moratalla, M. L., Veciana-Nogués, T., Bover-Cid, S., Garriga, M., Aymerich, T., Zanardi, E., & Vidal-Carou, M. C. (2008). Biogenic amines in traditional fermented sausages produced in selected European countries. *Food Chemistry, 107*(2), 912–921.
37. Lin, X., Han, P., Dong, S., & Li, H. (2015). Preparation and application of bacteriophage-loaded chitosan microspheres for controlling *Lactobacillus plantarum* contamination in bioethanol fermentation. *RSC Advances, 5*(85), 69886–69893.
38. Lücke, F. K. (1994). Fermented meat products. *Food Research International, 27*(3), 299–307.
39. Manassi, C. F., de Souza, S. S., de Souza Hassemer, G., Sartor, S., Lima, C. M. G., Miotto, M., & Verruck, S. (2022). Functional meat products: Trends in pro-, pre-, syn-, para- and post-biotic use. *Food Research International, 154*, 111035.
40. Marco, M. L., Heeney, D., Binda, S., Cifelli, C. J., Cotter, P. D., Foligné,

B., & Hutkins, R. (2017). Health benefits of fermented foods: Microbiota and beyond. *Current Opinion in Biotechnology, 44*, 94–102.

41. Martens, J. H., Barg, H., Warren, M. A., & Jahn, D. (2002). Microbial production of vitamin B12. *Applied Microbiology and Biotechnology, 58*, 275–285.

42. Martinez-Villaluenga, C., Peñas, E., Sidro, B., Ullate, M., Frías, J., & Vidal-Valverde, C. (2012). White cabbage fermentation improves ascorbigen content, antioxidant and nitric oxide production inhibitory activity in LPS-induced macrophages. *LWT–Food Science and Technology, 46*(1), 77–83.

43. Meisel, H. (2001). Bioactive peptides from milk proteins: A perspective for consumers and producers. *Australian Journal of Dairy Technology, 56*(2), 83–88.

44. Mills, S., Stanton, C., & Ross, R. P. (2009). Microbial production of bioactive: From fermented functional foods to probiotic mechanisms. *Australian Journal of Dairy Technology, 64*(1), 41–48.

45. Mıynat, M. E., Ören, İ., Özkan, E., & Argun, H. (2020). Sequential dark and photo-fermentative hydrogen gas production from agar-embedded molasses. *International Journal of Hydrogen Energy, 45*(60), 34730–34738.

46. Mukherjee, A., Breselge, S., Dimidi, E., Marco, M. L., & Cotter, P. D. (2024). Fermented foods and gastrointestinal health: Underlying mechanisms. *Journal of Functional Foods, 21*(4), 248–266.

47. Munekata, P. E., Pateiro, M., Zhang, W., Domínguez, R., Xing, L., Fierro, E. M., & Lorenzo, J. M. (2020). Autochthonous probiotics in meat products. *Microorganisms, 8*(11), 1833.

48. Narzary, Y., Brahma, J., Brahma, C., & Das, S. (2016). A study on indigenous fermented foods and beverages of Kokrajhar, Assam, India. *Journal of Ethnic Foods, 3*(4), 284–291.

49. Olasupo, N. A., Smith, S. I., & Akinsinde, K. A. (2002). Examination of the microbial status of selected indigenous fermented foods in Nigeria. *Journal of Food Safety, 22*(2), 85–93.

50. Ouoba, L. I. I., Nyanga-Koumou, C. A. G., Parkouda, C., Sawadogo, H., Kobawila, S. C., Keleke, S., & Sutherland, J. P. (2010). Genotypic diversity of lactic acid bacteria isolated from African traditional alkaline-fermented foods. *Journal of Applied Microbiology, 108*(6), 2019–2029.

51. Papavergou, E. J., Savvaidis, I. N., & Ambrosiadis, I. A. (2012). Levels of biogenic amines in retail market fermented meat products. *Food Chemistry, 135*(4), 2750–2755.

52. Park, Y. W., & Nam, M. S. (2015). Bioactive peptides in milk and dairy products: A review. *Korean Journal for Food Science of Animal Resources, 35*(6), 831–841.

53. Parkouda, C., Nielsen, D. S., Azokpota, P., Ouoba, L. I. I., Amoa-Awua, W. K., Thorsen, L., & Jakobsen, M. (2009). The microbiology of alkaline fermentation of indigenous seeds used as food condiments in Africa and Asia. *Critical Reviews in Microbiology, 35*(2), 139–156.

54. Pennacchia, C., Ercolini, D., Blaiotta, G., Pepe, O., Mauriello, G., & Villani, F. (2004). Selection of *Lactobacillus* strains from fermented sausages for their potential use as probiotics. *Meat Science, 67*(2), 309–317.

55. Punia, H., Tokas, J., Malik, A., Sangwan, S., Baloda, S., Singh, N., & Mor, V. S. (2020). Identification and detection of bioactive peptides in milk and dairy products: Remarks about agro-foods. *Molecules, 25*(15), 3328.

56. Rai, A. K., & Jeyaram, K. (2015). Health benefits of functional proteins in fermented foods. In *Health Benefits of Fermented Foods and Beverages* (pp. 455–474).

57. Ramos, C. L., & Schwan, R. F. (2017). Technological and nutritional aspects of indigenous Latin American fermented foods. *Current Opinion in Food Science, 13*, 97–102.

58. Ramos, C. L., Bressani, A. P., Batista, N. N., Martinez, S. J., Dias, D. R., & Schwan, R. F. (2023). Indigenous fermented foods: Nutritional and safety aspects. *Current Opinion in Food Science, 53*, 101075.

59. Rantsiou, K., Urso, R., Iacumin, L., Cantoni, C., Cattaneo, P., Comi, G., & Cocolin, L. (2005). Culture-dependent and-independent methods to investigate the microbial ecology of Italian fermented sausages. *Applied and Environmental Microbiology, 71*(4), 1977–1986.

60. Rigby, D. L., & Essletzbichler, J. (2006). Technological variety, technological change and a geography of production techniques. *Journal of Economic Geography, 6*(1), 45–70.

61. Roca, M., & Incze, K. (1990). Fermented sausages. *Food Reviews International, 6*(1), 91–118.

62. Rul, F., Béra-Maillet, C., Champomier-Vergès, M. C., El-Mecherfi, K. E., Foligné, B., Michalski, M. C., & Savary-Auzeloux, I. (2022). Underlying evidence for the health benefits of fermented foods in humans. *Food & Function, 13*(9), 4804–4824.

63. Sanders, M. E., & Marco, M. L. (2010). Food formats for effective delivery of probiotics. *Annual Review of Food Science and Technology, 1*, 65–85.

64. Şanlier, N., Gökcen, B. B., & Sezgin, A. C. (2019). Health benefits of fermented foods. *Critical Reviews in Food Science and Nutrition, 59*(3), 506–527.

65. Seol, K. H., Chang, O. K., Kim, M. K., Han, G. S., Jeong, S. G., Park, B. Y., & Ham, J. S. (2012). Production of bioactive peptides from milk. *Journal of Dairy Science and Biotechnology, 30*(1), 37–44.

66. Shiby, V. K., & Mishra, H. N. (2013). Fermented milks and milk products as functional foods—A review. *Critical Reviews in Food Science and Nutrition, 53*(5), 482–496.

67. Siciliano, R. A., Reale, A., Mazzeo, M. F., Morandi, S., Silvetti, T., & Brasca, M. (2021). Para probiotics: A new perspective for functional foods and nutraceuticals. *Nutrients, 13*(4), 1225.

68. Sivamaruthi, B. S., Kesika, P., & Chaiyasut, C. (2018). Impact of fermented foods on human cognitive function—A review of outcome of clinical trials. *Scientia Pharmaceutica, 86*(2), 22.

69. Stadnik, J., & Kęska, P. (2015). Meat and fermented meat products as a source of bioactive peptides. *Acta Scientiarum Polonorum Technologia Alimentaria, 14*(3), 181–190.

70. Stanton, C., Ross, R. P., Fitzgerald, G. F., & Van Sinderen, D. (2005). Fermented functional foods based on probiotics and their biogenic metabolites. *Current Opinion in Biotechnology, 16*(2), 198–203.

71. Sun, W., Shahrajabian, M. H., & Lin, M. (2022). Research progress of fermented functional foods and protein factory-microbial fermentation technology. *Fermentation, 8*(12), 688.

72. Suzzi, G., & Gardini, F. (2003). Biogenic amines in dry fermented sausages: A review. *International Journal of Food Microbiology, 88*(1), 41–54.

73. Syngai, G. G., Gopi, R., Bharali, R., Dey, S., Lakshmanan, G. A., & Ahmed, G. (2016). Probiotics—The versatile functional food ingredients. *Journal of Food Science and Technology, 53*, 921–933.

74. Takeyama, H., Kanamaru, A., Yoshino, Y., Kakuta, H., Kawamura, Y., & Matsunaga, T. (1997). Production of antioxidant vitamins, β-carotene, vitamin C, and vitamin E, by two-step culture of *Euglena gracilis* Z. *Biotechnology and Bioengineering, 53*(2), 185–190.

75. Tolonen, M., Taipale, M., Viander, B., Pihlava, J. M., Korhonen, H., & Ryhänen, E. L. (2002). Plant-derived biomolecules in fermented cabbage. *Journal of Agricultural and Food Chemistry, 50*(23), 6798–6803.

76. Torres, S., Verón, H., Contreras, L., & Isla, M. I. (2020). An overview

of plant-autochthonous microorganisms and fermented vegetable foods. *Food Science and Human Wellness, 9*(2), 112–123.

77. Tran, C. T., Nosworthy, N., Bilek, M. M., & McKenzie, D. R. (2015). Covalent immobilization of enzymes and yeast: Towards a continuous simultaneous saccharification and fermentation process for cellulosic ethanol. *Biomass and Bioenergy, 81*, 234–241.
78. Voidarou, C., Antoniadou, M., Rozos, G., Tzora, A., Skoufos, I., Varzakas, T., & Bezirtzoglou, E. (2020). Fermentative foods: Microbiology, biochemistry, potential human health benefits and public health issues. *Foods, 10*(1), 69.
79. Weaver, G. A., Krause, J. A., Miller, T. L., & Wolin, M. J. (1992). Cornstarch fermentation by the colonic microbial community yields more butyrate than does cabbage fiber fermentation. *The American Journal of Clinical Nutrition, 55*(1), 70–77.
80. Webb, A. R., DeCosta, B. R., & Holick, M. F. (1989). Sunlight regulates the cutaneous production of vitamin D3 by causing its photodegradation. *The Journal of Clinical Endocrinology & Metabolism, 68*(5), 882–887.
81. Yadav, H., Jain, S., Rastamanesh, R., Bomba, A., Catanzaro, R., & Marotta, F. (2011). Fermentation technology in the development of functional foods for human health: Where we should head. *Fermentation Technology, 1*, 1–2.

CHAPTER 4

BIOETHANOL PRODUCTION FROM MOLASSES BIOMASS

LEARNING OBJECTIVES

At the end of this chapter, readers will be able to:

1. Understand the composition and characteristics of different types of molasses used in ethanol production.
2. Explore the fermentation process of molasses, focusing on stages like feed preparation, yeast selection, and monitoring conditions.
3. Understand the efficiency and characteristics of Saccharomyces spp. strain in ethanol fermentation from molasses.
4. Know the strategies to enhance ethanol production from molasses using immobilized K. marxianus biocatalysts.
5. Understand Z.mobilis fermentation kinetics and productivity in high molasses concentrations.
6. Know the impact of additives on ethanol yield and productivity in molasses fermentation.

INTRODUCTORY EXAMPLE

Raízen, a prominent player in the bioethanol industry, exemplifies the utilization of molasses for ethanol production. As a joint venture between Cosan and Shell, Raízen capitalizes on molasses, a byproduct of sugarcane processing, to sustain one of the world's largest ethanol production operations. By employing advanced fermentation technologies and optimizing yeast strains, Raízen demonstrates the practicality and efficiency of molasses as a primary feedstock for ethanol. This strategic approach not only enhances production output but also underscores the economic and environmental benefits of utilizing agricultural residues for renewable energy solutions. Through their extensive operations, Raízen contributes significantly to meeting global energy demands sustainably while reducing environmental impact.

UNIT INTRODUCTION

Currently, a significant amount of research is focused on enhancing bioconversion strategies, investigation of the technological and economical feasibility, as well as the potential environmental consequences of above processes. Specifically, there has been ongoing research on producing ethanol from molasses for several years. The inquiry has primarily focused on two aspects: the addition of molasses and the utilization of heat-tolerant strains. These methods aim to improve both the rate of alcohol synthesis and the final ethanol concentration (Rasmey et al., 2018).

Cane molasses is a valuable by-product that is obtained as the final run-off syrup during the process of sugar manufacture. The substance is a thick, dark liquid that is left over after a process. Molasses contains a significant number of residual sugars, making it a viable substrate for large-scale ethanol production in industries. Molasses is fermented by yeast in the creation of commercial ethanol. To make ethanol from diluted cane molasses, some distilleries in India use an open fermentation system in a batch process (Farahani & Asoodar, 2017).

Despite India's position as the leading global producer of sugarcane and sugar, the production of bio-ethanol has not surpassed 1.5 billion liters per year, with a capacity utilization of approximately 60%. One possible reason for this could be the lack of temperature control during fermentation in distilleries located in tropical regions of India, especially during the hot summer season. The activity of the fermenting organisms is negatively impacted by these high temperatures, which also enhances the toxic effect of ethanol. As a result, fermentation efficiency decreases and the fermentation process ends prematurely (Jones et al., 1981).

4.1. TYPES OF MOLASSES

Learning Objectives:

- Understand the types and compositions of various molasses used in production.

The Association of American Feed Control Officials (AAFCO, 1982) provides a description of the various varieties of molasses and their composition in Table 4.1. Cane molasses is a by-product that is obtained during the process of manufacturing or refining sucrose from sugarcane. The product has a minimum sugar content of 46%. Beet molasses has a high concentration of total sugars, with a density of approximately 79.5° Brix. The juiceis partially dried and obtained from the process of making dried citrus pulp has a minimum of 45% total sugar and a Brix density of roughly 71.0. A by-product of the pressed wood manufacturing process is a hemicellulose extract (Palmonari et al., 2020).

Without the use of alkalis, acids, or salts, wood is treated at high temperatures and pressures to produce the concentrated soluble substance. The substance includes hexose and pentose sugars, with a minimum total carbohydrate content of 55%. A by-product of the extraction of dextrose from corn or grain sorghum is starch molasses.

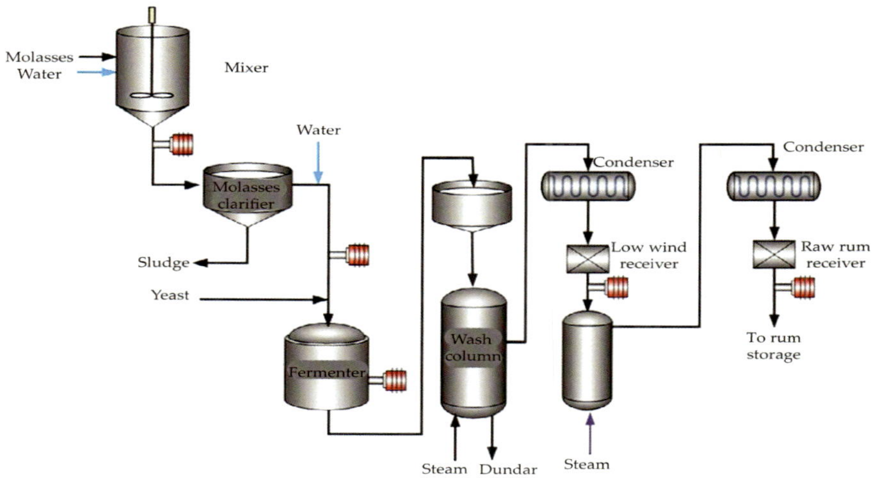

Figure 4.1. *Illustration of ethanol production from molasses.*

Source: Anand Parkash, Creative Commons License.

This extraction involves the hydrolysis of starch using enzymes or acid. The composition includes approximately 73% total solids and 43% reducing sugars. Based on the data available, it appears that cane molasses accounts for

60% of the total U.S. molasses supply, while beet molasses makes up 32%. Starch molasses and citrus molasses, on the other hand, contribute 7% and 1%, respectively. There are significant limitations for the production of citrus molasses, starch molasses, and hemicellulose extract (Hatfield et al., 1998) (Figure 4.1).

Table 4.1. *Composition of Various Types of Molasses*

	Type of Molasses				
Item	**Cane**	**Beet**	**Extract**	**Citrus**	**Starch**
Brix	79.5	79.5	65.0	71.0	78.0
Specific gravity	1.41	0.41	1.32	1.36	1.40
Total solids (%)	75.0	77.0	65.0	65.0	73.0
Crude protein (%)	3.0	6.0	0.5	4.0	0.4
Total sugars (%)	46.0	48.0	55.0	45.0	50.0
Total fat (%)	0.0	0.0	0.5	0.2	0.0
Nitrogen-free extract (%)	63.0	62.0	55.0	55.0	65.0
Ash (%)	8.1	8.7	5.0	6.0	6.0
Total fiber (%)	0.0	0.0	0.5	0.0	0.0
Phosphorus (%)	0.08	0.03	0.05	0.15	0.2
Calcium (%)	0.8	0.2	0.8	1.3	0.1
Sodium (%)	0.2	1.0	–	0.3	2.5
Potassium (%)	2.4	4.7	0.04	0.1	0.02
Sulfur (%)	0.5	0.5	–	0.17	0.05
Chlorine (%)	1.4	0.9	–	0.07	3.0
Swine (ME)	2,343	2,320	2,231	2,264	–

Source: Velusamy Senthilkumar, Creative Commons License.

4.2. GENERAL PROCESS FOR THE PRODUCING ETHANOL FROM MOLASSES

Learning Objectives:

- Understand the process stages involved in ethanol production from molasses.

Production of ethanol in distilleries consists of three primary stages: feed preparation, fermentation, and distillation (Figure 4.2). To get a solution with the right amount of sugar, molasses and water are combined. Sulfuric acid can be added to change the pH if needed. After being transferred to the fermentation tank, the diluted molasses solution is introduced with a 10% yeast seed culture. Subsequently, the combination undergoes fermentation in an anaerobic environment by continuous monitoring of its pH and temperature (Castañeda & Cortez, 2017).

Due to the exothermic nature of the reaction, it is necessary to cool the fermenter in order to keep the reaction temperature at a steady 25°C. The fermentation process usually requires 48 to 80 hours, resulting in a broth with an ethanol content of 6 to 8%. After the fermentation process, the yeast is separated through settling, and the broth without cells is collected for distillation. Six to nine fermenters are often used by Indian distilleries to ensure a consistent supply of alcohol. (Atiyeh & Duvnjak, 2002). The fermentation process can be conducted either in batch or continuous mode.

Distilleries often choose continuous fermentation due to its higher efficiency, simple operation, and significant reduction in water consumption. Continuous fermentation is 89 to 90% efficient, compared to the 80 to 84% efficiency of batch mode. After being heated to about 90°C, the fermented broth is sent to the analyzer column's degasifying section. Any trapped gases, like CO_2, are successfully removed from the liquor by the bubble cap fractionating column. The liquor is then subjected to steam heating and fractionation, which leads to a 40% yield of alcohol. The output at the bottom of the analyzer column is referred as the effluent, sometimes known as the spent wash (Bouallagui et al., 2013). The rectifying column receives the alcohol vapors from the analyzer column and uses a reflux action to collect highly pure alcohol (95–99%).

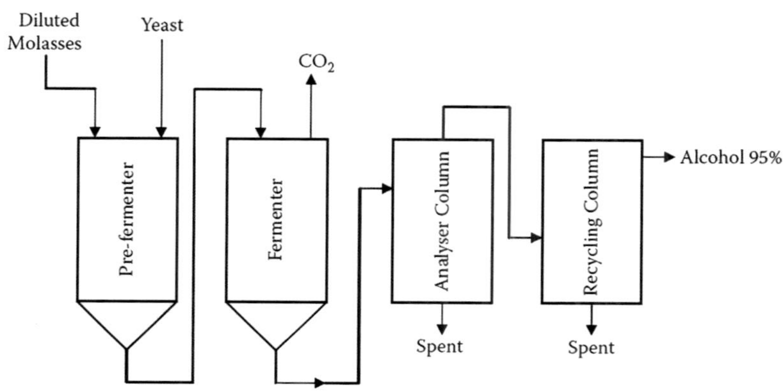

Figure 4.2. *Illustration of the manufacturing process of ethanol from molasses.*

Source: Paramasamy Gunasekaran, Creative Commons License.

4.3. FERMENTATION OF MOLASSES BY *SACCHAROMYCES* SPP

Learning Objectives:

- Understand the role of *Saccharomyces* spp. strains in ethanol fermentation from molasses.

Yeast strains from the *Saccharomyces* spp. family are frequently used in the production of ethanol derived from cane molasses. It is crucial to have yeast strains that possess high specific ethanol productivity and can tolerate the concentrations of substrate and product at the typical temperatures in the surrounding areas. This is vital for ensuring an efficient process. It becomes essential to have osmotolerant yeast when using raw materials like cane that contain a lot of salt and various forms of blackstrap molasses (Ngwenya et al., 2012).

Flocculation is another desirable characteristic that facilitates continuous fermentation by allowing yeast cells to be retained in tower reactors and making cell recovery in batch fermentation easier. Various yeast strains have undergone testing to evaluate their effectiveness in ethanol fermentation only a few of them being utilized for large-scale ethanol production (Table 4.2). There is little information available about the relative performance of different yeasts on high-salt molasses (Arshad et al., 2017).

Ragav researched the impact of a modified strain of *Saccharomyces uvarum* in batch fermentation of sugarcane molasses. The outcomes were then contrasted with those from *S. cerevisiae* Y-10, a strain which is tolerant of ethanol and substrate, and *S. uvarum* ATCC 26602, a strain used for normal brewing. In a study carried by Comberbach & Bu'Lock (1984), the *S. uvarum* strain 17 was used to achieve a fast and effective continuous fermentation process, converting glucose into ethanol. Various strains of *S. cerevisiae* were studied to determine their capacity for producing ethanol in molasses containing high amount of sugar. The focus was to identify strains with favorable fermentation traits (Nahvi et al., 2002).

In the concentrated molasses (containing 40% sugars), four strains—Isolate 3B, *S. cerevisiae* MTCC 174, *S. cerevisiae* HAU11, and *S. cerevisiae* MTCC 172, produced ethanol with great efficiency (71.0, 67.0, 66.7, and 61.5%, respectively). The yeast strains' viability was quite high in the diluted molasses but sharply declined as the medium's sugar concentration increased and as the incubation period lengthened (El-Refai et al., 1992).

Table 4.2. The Relative Efficiency of the Yield Strains Used in the Commercial Production of Ethanol

Yeast Strain	**Ethanol/Ton of Molasses (Gallons)**	**Fermentation Efficiency (%)**
ATCC 4132	73	93
CBS 237	70	90
Y 7494	67	86
UCD 505	65	83
UCD 595	63	81
ATCC 26603	63	81
DADY	60	77
BAKER	60	77
ATCC 26602	48	62
NCYC 90	44	57
Y 2034	43	55
CBS 1235	27	35

Source: Paramasamy Gunasekaran, Creative Commons License.

The four superior strains (3B, *S. cerevisiae* MTCC 174, *S. cerevisiae* MTCC 172, and *S. cerevisiae* HAU-11) exhibited cell viability ranging from 57 to 71% in molasses with a sugar concentration of 35 to 40%. Following UV treatment, thorough screening, and molasses medium fermentation modification, thermotolerant *S. cerevisiae* MT15 was isolated. The mutation changed the behavior and metabolite-producing ability of the culture (Khoja et al., 2015).

When cultivated on molasses (which contained 15% sugars, w/v), this mutant produced the highest volumetric alcohol production of 72 g/l at 40°C. This yield was greater than molasses or glucose using the well-researched *Kluyveromyces marxianus* IMB-3. The organism showed a quick fermentation capacity at temperatures up to 40°C, with considerably greater substrate consumption characteristics ($P < 0.05$) when compared to the wild strain and five other strains of *K. marxianus*.

The mutant showed a significant improvement compared to its wild parent in reference to ethanol productivity, product yield, and specific ethanol yield. The ethanol productivity reached 7.2 g/l/h, the product yield was 0.44 g ethanol/g substrate consumed, and specific ethanol yield was 19.0 g ethanol/g cells. There was a noticeable link between the enhanced ethanol production and the level of extracellular and intracellular invertase activity. The mutant strain had a much better capacity to produce ethanol than the parental strain, with both volumetric and product yields showing increases ($P < 0.05$). The cell system protected thermal inactivation during creation, according to a thermodynamic study (Abubaker et al., 2012).

4.3.1. Ethanol Fermentation by the Cell Recycle System

The recycling ratio and dilution rate were discovered to influence the productivity of *S. cerevisiae*'s continuous cell recycle fermentation. The study revealed that the productivity of ethanol showed an upward trend as the dilution rate increased from 0.2/h to 0.3/h. However, it was observed that the ethanol productivity declined when the dilution rate exceeded this threshold. The cell washout likely occurred as a result of the system being operated at higher dilution rates (Cysewski & Wilke, 1977).

The pilot recycling circulation culture reached its peak output of 20.61 ml/l/h when the dilution rate was 0.3/h, and the recycling ratio was 9. As the dilution rate increased, there was a noticeable decrease in the concentration of cells in the fermenter. There was an apparent rise in the up-flow rate inside the sedimentation vessel when the dilution rate went over 0.3/h, which resulted in a drop in cell concentration (Lee et al., 1983) (Figure 4.3).

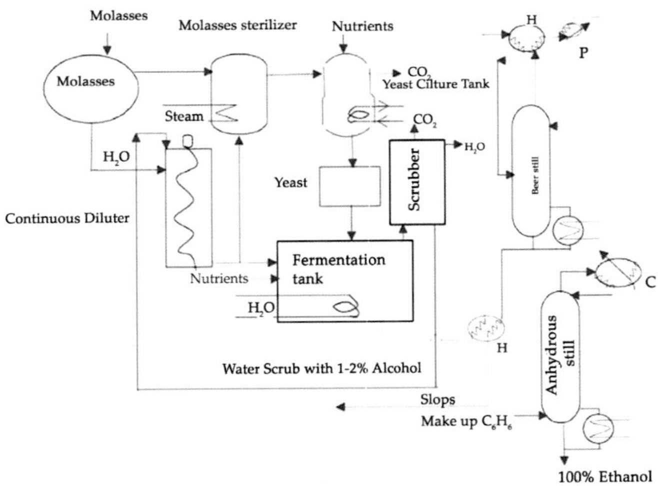

Figure 4.3. *Illustration of ethanol production from bioresources.*

Source: Paramasamy Gunasekaran, Creative Commons License.

There was a noticeable rise in the concentration of cells in the fermenter when the recycling ration was increased. For a second fermentation cycle, more unused medium was added to the primary fermenter. When the circulating ratio exceeded 9.0, there was a remarkable uniformity in the concentration. The concentration of cells in the separating vessel and the fermenter was almost the same (Thani et al., 2017).

The flow condition in the fermenter and separation vessel is influenced by the feed rate and circulating ratio. When choosing strains for ethanol fermentation, it is important to consider their growth rate and ability to separate well at high ethanol concentrations (Bo et al., 2006).

4.4. FERMENTATION OF MOLASSES USING THE THERMO-TOLERANT YEAST *K. MARXIANUS*

Learning Objectives:

- Understand the fermentation capabilities and optimization strategies of thermotolerant *K. marxianus*.

One of the main drawbacks of fermentation during molasses fermentation is an exothermic process. Multiple strains of the thermotolerant yeast *K. marxianus* have demonstrated effective resolution of this issue. Studies have been demonstrated that the thermotolerant yeast strain *K. marxianus* is capable of synthesizing ethanol at temperatures up to 45°C, by utilizing various types of complex and simple carbohydrate substrates (Pimpakan et al., 2012).

Furthermore, it has been proven that yeast can generate ethanol from diluted, supplemented molasses. A yeast cell preparation that is immobilized can be utilized as the biocatalyst in various fermentation processes. The highest level of ethanol generation by *K. marxianus* IMB3 was seen when using molasses at a concentration of 23% (v/v). At this concentration, a solution containing 7.4% ethanol by volume was generated, which accounted for 84% of the expected maximum yield (Gough et al., 1998).

The ethanol production rate was 1 g/l/h. When the molasses concentration exceeded 23% (v/v), both biomass and the maximum ethanol concentration decreased. No ethanol was created when the molasses concentration reached 44% (v/v). When sucrose was added in increasing amounts from 140 to 180 g/l, the concentration of ethanol was decreased. This drop was similar to the decline observed in the molasses dilution studies, where the total sugar concentration was matched (Wu et al., 2016).

A research investigation examining the impact of four supplements, namely nitrogen, magnesium, potassium, and linseed oil, on the rate of fermentation and the ultimate concentration of ethanol, revealed a significant enhancement in both the rate of ethanol production (4.8 g/l/h) and the concentration of ethanol

(8.5% v/v). The impacts on the biomass concentration and the specific ethanol yield could not be distinguished due to the inability to determine the biomass concentration. Research has indicated that the co-purposing of magnesium sulfate and linseed oil can positively influence rate of ethanol production (Chang et al., 2014).

Practice Problem

In the bioethanol industry, molasses serves as a valuable raw material for ethanol production due to its high sugar content derived from sugarcane processing. Companies like CalTech, a leading ethanol producer, utilize molasses effectively in their operations. CalTech, a joint venture between Cosan and Shell, relies on molasses as a primary feedstock for ethanol production. Discuss the strategies employed by CalTech.

Solutions to Practice Problem

CalTech employs several strategies to maximize the utilization of molasses for ethanol production. Firstly, they optimize fermentation processes using advanced yeast strains that are adept at fermenting the sugars present in molasses efficiently. This ensures higher ethanol yields per unit of molasses used. Secondly, CalTech implements a robust quality control system to monitor the composition of molasses, ensuring consistent fermentation performance and ethanol quality. Thirdly, they have invested in technology and infrastructure for continuous fermentation processes, which allows for efficient and continuous production of ethanol from molasses. These strategies collectively enhance CalTech capacity to produce ethanol sustainably and competitively in the global market, highlighting the economic and environmental benefits of utilizing molasses as a feedstock for bioethanol production.

4.4.1. Strategies to Improve Ethanol Production by K. marxianus

An immobilization technique was used to protect a heat-resistant strain of *K. marxianus* IMB3 within calcium alginate matrices. An investigation was done to assess the biocatalyst's capacity to generate ethanol from cane molasses sourced from various countries including Guatemala, Honduras, Senegal, Guyana, and the Philippines. For each case, the molasses was diluted to achieve a sugar concentration of 140 g/l. The fermentations were then conducted in a batch-fed mode at a temperature of 45°C. Within the initial 24 hours, range of 43 to 57 g/l was achieved as the highest ethanol concentrations, with the most favorable production occurring using molasses from Honduras (Sousa et al., 2018). The ethanol production gradually decreased during the 120 hours as the fermentations were fed every 24 hours, resulting in concentrations ranging from 20 to 36 g/l. Additionally, the fermentations that contained the molasses from Guyana maintained the highest ethanol productivity. After each round of fermentation, a subsequent feeding at 120 hours resulted in another 48 hours of continued production, leading to a peak in ethanol production. The concentrations of ethanol ranged from 25 to 52 g/l during this period. Nevertheless, extending the duration between the refeeding during this phase of fermentation harmed the efficiency of the biocatalyst (Moreno et al., 2012).

The study investigated the potential of utilizing tamarind byproducts, including pulp, seeds, husk, fruit, and effluent from tartaric acid extraction, as additives to enhance alcohol production from cane molasses. Adding a small amount of these substances increased the speed at which ethanol was produced in batch fermentations. The ethanol production from the molasses' reducing sugars was significantly enhanced by tamarind fruit, resulting in an increase of 6.5 to 9.7% (w/v).

Overall, incorporating tamarind into the fermentation process resulted in a significant increase of over 40% in ethanol production when using greater quantities of cane molasses sugar. The aqueous tamarind effluent may be directly fermented to produce 3.25% (w/v) ethanol. This indicates that it could be potentially used as a diluent in molasses fermentations. The study utilized fresh, chilled, and lignin-free brewer's waste grains (BSG) to enhance the alcoholic fermentation of molasses by yeast (Zafar & Owais, 2006).

A 12% w/v glucose solution was utilized for immobilizing cells on fresh brewer's spent grain (BSG), with and without nutrients. For immobilizing cells on defrosted BSG, a glucose solution without nutrients was applied. On the other hand, for immobilizing cells on delignified BSG, a glucose solution with nutrients was used. The immobilized biocatalysts underwent multiple fermentation batches in molasses with initial Baume densities of 7, 10 and 12, at temperatures of 30 and 20°C, without the addition of further nutrients. The chilled BSG immobilized biocatalyst was exclusively utilized for consecutive batches of molasses with an initial Baume density of 7, without any added nutrients, at temperatures of 30 and 20°C (Yuan et al., 2013).

Following immobilization, the population of immobilized microorganisms reached 109 cells per gram of support material for all the immobilized biocatalysts. Use of fresh BSG immobilized biocatalyst produced better ethanol productivity during the initial stages of molasses fermentation, lower ultimate Baume densities, and greater fermentation rates when compared to alternative biocatalysts. Baume densities are in the range of 7 to 12.

In the molasses fermentation system, the defrosted BSG immobilized biocatalyst demonstrated adaptation, with its kinetic characteristics resembling those found in the fresh BSG immobilized biocatalyst batch by batch (Saini et al., 2017).

Hence, utilizing fresh or defrosted BSG as yeast supports shows potential for successful scale-up operations. The alcoholic fermentation of molasses was investigated using *S. cerevisiae* immobilized on orange peel pieces, with temperatures ranging from 30 to 15°C. The fermentation durations in all instances were short (5–15 hours) and the rates at which ethanol was produced were high (150.6 grams per liter per day), indicating excellent stability and appropriateness of the biocatalyst for commercial use. The study found that the proposed biocatalyst produced significant levels of volatile by-products at all temperatures tested. This shows that the biocatalyst may be able to improve quality and productivity in fermented food applications.

Studies have indicated that when alginate is used as the immobilizing matrix, its structural integrity deteriorates when the matrix is used for extended periods. Therefore, it becomes essential to add calcium to the media/reactor feedsto maintain its stability. In continuous or semi-continuous operations, polyvinyl alcohol cryogel (PVAC) beads were evaluated as a substitute immobilization matrix for alginate for the immobilized cells.

Using a fed-batch mode, the biocatalyst, which was immobilized in alginate, achieved ethanol concentrations of up to 57 g/l after 48 hours. This was achieved from a sugar concentration of 140 g/l, resulting in an 80% theoretical yield. After a further 425 hours of re-feeding, the ethanol concentration in the fermentations containing the alginate-based biocatalyst was significantly reduced to 20 g/l.

Throughout the extended duration of 60 to 500 hours, the ethanol content was maintained at a consistently low level. The mean ethanol concentration obtained over the 500-hour duration was determined to be 21 g/l, which accounted for 29% of the highest possible output. With peak concentrations of 52 to 53 g/l (or 73% of the theoretical yield), the PVAC-immobilized biocatalyst was able to convert molasses to ethanol in less than 72 hours (Gough et al., 1998).

The mean ethanol concentration generated throughout a 600-hour timeframe was determined to be 45 grams per liter, equivalent to 63% of the potential yield. The causes for this significant disparity in productivity, especially during extended running durations, are yet unidentified. Conversely, initial findings indicate that the biocatalyst immobilized in PVAC outlives the immobilized alginate-based system in terms of viability.

A study was carried out to investigate how different molasses sugar concentrations affected the generation of ethanol. This study focused on using alginate-immobilized *K. marxianus* in a continuous-flow bioreactor. Using sugar concentrations of 140 g/l over 10 hours produced the greatest ethanol concentrations. Within 48 hours, the ethanol concentrations gradually dropped to lower levels. Over this time, there was a significant decrease in the number of yeast cells within the immobilization matrix. When molasses concentrations were reduced, production of ethanol showed little variation (Goshima et al., 2013).

A study was carried out to examine the influence of residence time on the production of an ethanol in a bioreactor that operates with a continuous flow. Through experimentation, it was shown that a residence duration of 0.66 hours resulted in the best volumetric production when the molasses sugar concentration was fixed at 120 g/l.

4.5. *ZYMOMONAS MOBILIS* POTENTIAL TO PRODUCE ETHANOL FROM MOLASSES

Learning Objectives:

- Understand the advantages of *Zymomonas mobilis* over yeast in ethanol production from molasses.
- Explore fermentation kinetics and optimization strategies for *Z. mobilis* using molasses.

Numerous strategies for enhancing the ethanol fermentation processes have been developed in response to the increased demand for alcohol. There are several benefits that the bacterium *Z. mobilis* has over yeast fermentation in the hunt for an effective ethanol-producing organism. These include (i) increased yield of ethanol and sugar uptake; (ii) reduced biomass production; (iii) increased tolerance to ethanol; (iv) no requirement for regulated oxygen addition during fermentation; and (v) susceptibility to genetic modification. Only glucose, fructose, and sucrose can be used by *Z. mobilis* strains with high fermentation efficiency (Cazetta et al., 2007) (Figure 4.4).

On the other hand, the sucrose yields are relatively low because of the production of by-products like sorbitol and levan. Various efforts have been undertaken to explore ethanol fermentation using commonly available substrates like cane and beet molasses. Nevertheless, production of ethanol from molasses is hindered by the formation of by-products and the presence of inorganic ions. It has been reported that mutant strains have been chosen for their ability to efficiently hydrolyze beet molasses and ferment cane (Panesar et al., 2006).

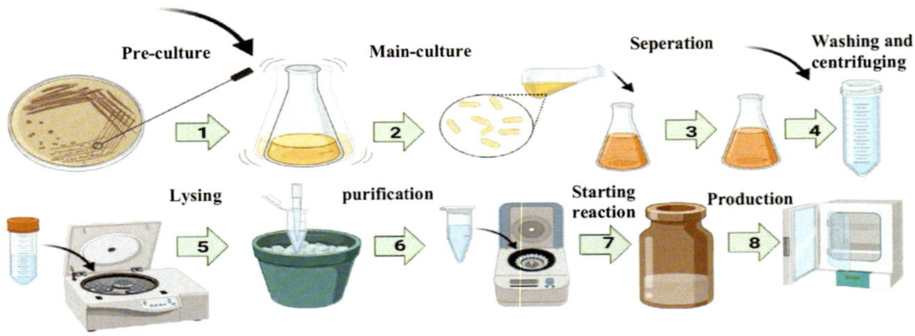

Figure 4.4. *Illustration of ethanol production using the Zymomonas mobilis.*

Source: Adelaide Braga, Creative Commons License.

4.5.1. Z. mobilis Adapted for Cane Molasses Fermentation

Table 4.3 displays the characteristics of *Z. mobilis* ZM4A's fermentation of molasses (20% w/v) at 30°C. With 91.2% substrate consumption, the maximum ethanol output of 0.47 g/g was attained (Jain & Singh, 1994). It has been reported that molasses can ferment when partially supplemented with mineral salts or when *Z. mobilis* yeast extract is used. During the fermentation process without using minerals, a maximum of 39.4 g/l final ethanol concentration and 91.3 g/l substrate utilization were noted after 24 hours (Jain & Singh, 1994).

As a result, the molasses medium offered some buffering capability without requiring the addition of supplements because the pH was left unchanged. Without the additives, the molasses (90 g/l sugar content) had an ethanol yield of 0.48 g/g. According to Park and Baratti, *Z. mobilis* produced more ethanol when 0.5 g/l of magnesium sulfate was added to the sugar beet molasses medium (Doelle et al., 1990).

Table 4.3. *Ethanol Production by Z. mobilis from Molasses Medium*

Overall Parameters	
Initial sugar (g/l)	110.0
Residual sugar (g/l)	9.6
Biomass (g/l)	1.6
Ethanol (g/l)	47.0
Substrate utilized (g/l)	91.2
Ethanol yield (g/l)	0.47
Biomass yield (g/l)	0.016

Fermentation efficiency (%)	92.0
Fermentation time (h)	24.0

Source: Paramasamy Gunasekaran, Creative Commons License.

4.5.2. Z. mobilis Fermentation Kinetics at High Molasses Concentration

Batch fermentation kinetics were conducted at greater molasses concentrations since *Z. mobilis* proved effective in fermenting 20% (w/v) of molasses. The batch fermentation utilizing molasses (110 g/l sugar concentration) yielded the highest ethanol production (47 g/l) with a maximum substrate consumption of 91.0% w/v. In addition, the specific substrate absorption (7.50 to 7.74 g/g/h), specific ethanol productivity (3.12 to 3.56 g/g/h), specific growth rate (0.128 to 0.137 μ/h), and fermentation efficiency (76.3 to 92.0) were all greater than those of the 40% molasses medium (Favela et al., 1986).

The presence of 40% molasses concentration, equivalent to 200 g/l of sugar, had a detrimental effect on cell growth. The explanation for this phenomenon lies in the interaction between ethanol inhibition and the impact of elevated osmotic pressure resulting from higher molasses concentration. Research has found that when comparing the fermentation of sucrose and molasses by *Z. mobilis* ZM4, sucrose is more effectively converted to ethanol at high concentrations (200 g/l). In fact, sucrose fermentation yielded an impressive 88.0 g/l of ethanol (Park & Baratti, 1992). On the other hand, the presence of inorganic ions has a notable inhibitory effect on the fermentation of molasses medium containing 200 g/l sugars.

4.5.3. Z. mobilis Continuous Fermentation of Diluted Molasses

Savvides conducted research on *Z. mobilis* CP4 mutants and in a Z recombinant *Z. mobilis* strains to explore their potential for ethanol production using molasses. Under the conditions of a sucrose medium, the production of ethanol is closely aligned with the steady-state biomass. The ethanol production rate of 43 g/1 was consistently maintained by the wild strain, suc40/pDS3154-inaZ and suc40. There was a notable 80% decrease in ethanol production after that. The growing medium used was sugar beet molasses, and the ethanol yields of both strains (suc40/pDS3154-inaZ and suc40) were identical (Silbir et al., 2014).

The mutant strain displayed rapid growth and remarkable stability in a medium with a high concentration of sugar beet molasses. An analysis of the strains revealed the presence of significant amounts of long-chain unsaturated fatty acids, particularly vaccenic acid (18:1). Interestingly, the mutant strain exhibited even higher levels of this fatty acid, reaching approximately 80%. In a study conducted by (Carey & Ingram, 1983), it was proposed that the organism's ability to thrive in high ethanol concentrations could be attributed by the presence of vaccenic acid. This is because long-chain unsaturated fatty acids balance out the disruptive effect of ethanol on the membrane structure (Behera et al., 2012).

When different *Z. mobilis* isolates were fermenting in molasses, the pH range of 5.0 to 5.6 was found optimal for the formation of ethanol. When *Z. mobilis* 10988 and these isolates were compared for ethanol production, the study discovered that the isolates produced noticeably less ethanol. Temperature's influence on ethanol generation was observed throughout the fermentation process. Specifically, when the temperature was maintained at 32°C, ethanol production increased significantly, reaching a range of 46 to 50 g/l. However, when the temperature exceeded 34°C, production of ethanol was greatly hindered, along with the growth of biomass (Huertaz et al., 1991).

In experiments with varying molasses concentrations, observed that the yeast strain yielded higher amounts of ethanol at higher sugar concentrations (25 g/l). On the other hand, *Z. mobilis* showed a high ethanol production with a maximum theoretical yield at lower sugar concentrations (23 g/l). According to earlier studies, *Z. mobilis* strains can make ethanol in synthetic media at high sugar concentrations. The observed low efficacy could be due to the presence of certain chemicals in the molasses that prevent growth and the production of ethanol.

SUMMARY

- The chapter explores different aspects of ethanol production using molasses as a substrate. It explores various molasses options and provides an overview of the typical procedure for producing ethanol.
- Some of the specific topics covered in this area include the fermentation of molasses by *Saccharomyces* spp., ethanol fermentation using the cell recycle system, and the utilization of thermo-tolerant yeast *K. marxianus* for molasses fermentation. Exploring strategies to enhance ethanol production by *K. marxianus* is also a focus of the study.
- This chapter also investigates the potential applications of *Zymomonas mobilis* in the molasses-to-ethanol conversion process. It provides a comprehensive analysis of its ability to adapt to cane molasses fermentation, the fermentation kinetics when dealing with high molasses concentrations, and the continuous fermentation of diluted molasses by *Z. mobilis*.

REVIEW QUESTIONS

1. Describe ethanol production from molasses, including key steps and considerations.
2. Compare Saccharomyces spp. and Kluyveromyces marxianus in molasses fermentation.
3. Explain the cell recycle system's role in improving ethanol fermentation efficiency.
4. Discuss Zymomonas mobilis' suitability and characteristics for ethanol production.
5. Types of molasses used in bioethanol production.
6. Strategies to enhance ethanol production by Kluyveromyces marxianus.
7. Fermentation kinetics of Zymomonas mobilis with high molasses concentrations.
8. Compare batch, fed-batch, and continuous fermentation for ethanol production.
9. Impact of yeast strain choice on ethanol production efficiency and economics.

MULTIPLE CHOICE QUESTIONS

1. **What is the primary microorganism used for fermenting molasses into ethanol?**
   a. Escherichia coli
   b. Saccharomyces spp.
   c. Lactobacillus spp.
   d. Bacillus subtilis
2. **Which yeast species is known for its thermo-tolerant properties, making it suitable for ethanol production in tropical climates?**
   a. Saccharomyces cerevisiae
   b. Candida tropicalis
   c. Kluyveromyces marxianus
   d. Pichia pastoris
3. **What is a significant advantage of using Zymomonas mobilis for ethanol production from molasses?**
   a. High tolerance to high ethanol concentrations
   b. Ability to ferment cellulose directly
   c. Low sensitivity to pH fluctuations
   d. Production of high-value co-products
4. **Which process involves feeding diluted molasses continuously into a bioreactor while withdrawing fermented broth simultaneously?**
   a. Batch fermentation
   b. Fed-batch fermentation
   c. Continuous fermentation
   d. Semi-continuous fermentation
5. **Which system improves ethanol yield by recycling cells back into the fermentation process?**
   a. Cell recycling system
   b. Batch fermentation
   c. Continuous fermentation
   d. Fed-batch fermentation

Answers to Multiple Questions

1. (b); 2. (c); 3. (a); 4. (c); 5. (a)

REFERENCES

1. Abubaker, H. O., Sulieman, A. M. E., & Elamin, H. B. (2012). Utilization of *Schizosaccharomyces pombe* for production of ethanol from cane molasses. *Journal of Microbiol Research, 2*, 36–40.
2. Arshad, M., Hussain, T., Iqbal, M., & Abbas, M. (2017). Enhanced ethanol production at commercial scale from molasses using high gravity technology by mutant *S. cerevisiae. Journal of Biotechnology, 48*, 403–409.
3. Atiyeh, H., & Duvnjak, Z. (2002). Production of fructose and ethanol from sugar beet molasses using *Saccharomyces cerevisiae* ATCC 36858. *Biotechnology Progress, 18*(2), 234–239.
4. Behera, S., Mohanty, R. C., & Ray, R. C. (2012). Ethanol fermentation of sugarcane molasses by *Zymomonas mobilis* MTCC 92 immobilized in *Luffa cylindrica* L. sponge discs and Ca-alginate matrices. *Biotechnology Progress, 43*, 1499–1507.
5. Bo, W. A. N. G., Xu-Meng, G. E., Ning, L. I., & Feng-Wu, B. A. I. (2006). Continuous ethanol fermentation coupled with recycling of yeast flocs. *Chinese Journal of Biotechnology, 22*(5), 816–821.
6. Bouallagui, H., Touhami, Y., Hanafi, N., Ghariani, A., & Hamdi, M. (2013). Performances comparison between three technologies for continuous ethanol production from molasses. *Biomass and Bioenergy, 48*, 25–32.
7. Carey, V. C., Walia, S. K., & Ingram, L. O. (1983). Expression of a lactose transposon (Tn 951) in *Zymomonas mobilis. Journal of Bacteriology, 46*(5), 1163–1168.
8. Castañeda-Ayarza, J. A., & Cortez, L. A. B. (2017). Final and B molasses for fuel ethanol production and some market implications. *Renewable and Sustainable Energy Reviews, 70*, 1059–1065.
9. Cazetta, M. L., Celligoi, M. A. P. C., Buzato, J. B., & Scarmino, I. S. (2007). Fermentation of molasses by *Zymomonas mobilis*: Effects of temperature and sugar concentration on ethanol production. *Bioresource Technology, 98*(15), 2824–2828.
10. Chang, J. J., Ho, C. Y., Mao, C. T., Barham, N., Huang, Y. R., Ho, F. J., & Huang, C. C. (2014). A thermo-and toxin-tolerant kefir yeast for biorefinery and biofuel production. *Applied Energy, 132*, 465–474.
11. Cysewski, G. R., & Wilke, C. R. (1977). Rapid ethanol fermentations using vacuum and cell recycle. *Biotechnology and Bioengineering, 19*(8), 1125–1143.

12. Doelle, M. B., & Doelle, H. W. (1990). Sugar-cane molasses fermentation by *Zymomonas mobilis*. *Applied Microbiology and Biotechnology, 33*, 31–35.
13. El-Refai, A. H., El-Abyad, M. S., El-Diwany, A. I., Sallam, L. A., & Allam, R. F. (1992). Some physiological parameters for ethanol production from beet molasses by *Saccharomyces cerevisiae* Y-7. *Bioresource Technology, 42*(3), 183–189.
14. Farahani, S. S., & Asoodar, M. A. (2017). Life cycle environmental impacts of bioethanol production from sugarcane molasses in Iran. *Environmental Science and Pollution Research, 24*(28), 22547–22556.
15. Favela-Torres, E., Allais, J. J., & Baratti, J. (1986). Kinetics of batch fermentations for ethanol production with *Zymomonas mobilis* growing on Jerusalem artichoke juice. *Biotechnology and Bioengineering, 28*(6), 850–856.
16. Goshima, T., Tsuji, M., Inoue, H., Yano, S., Hoshino, T., & Matsushika, A. (2013). Bioethanol production from lignocellulosic biomass by a novel *Kluyveromyces marxianus* strain. *Bioscience, Biotechnology, and Biochemistry, 77*(7), 1505–1510.
17. Gough, L., & Grace, J. B. (1998). Herbivore effects on plant species density at varying productivity levels. *Ecology, 79*(5), 1586–1594.
18. Gough, S., Barron, N., Zubov, A. L., Lozinsky, V. I., & McHale, A. P. (1998). Production of ethanol from molasses at 45°C using *Kluyveromyces marxianus* IMB3 immobilized in calcium alginate gels and poly (vinyl alcohol) cryogel. *Bioprocess Engineering, 19*, 87–90.
19. Hatfield, P. G., Hopkins, J. A., Ramsey, W. S., & Gilmore, A. (1998). Effects of level of protein and type of molasses on digesta kinetics and blood metabolites in sheep. *Small Ruminant Research, 28*(2), 161–170.
20. Huertaz-Díaz, H., Cacho, C. L., & Bernard, L. (1991). Fermentation of sugarcane juice and blackstrap molasses by *Zymomonas mobilis*. *Journal of Agriculture University of Puerto Rico, 75*(1), 43–50.
21. Jin, M., Gunawan, C., Uppugundla, N., Balan, V., & Dale, B. E. (2012). A novel integrated biological process for cellulosic ethanol production featuring high ethanol productivity, enzyme recycling and yeast cells reuse. *Energy & Environmental Science, 5*(5), 7168–7175.
22. Jones, R. P., Pamment, N., & Greenfield, P. F. (1981). Alcohol fermentation by yeasts: The effect of environmental and other variables. *Process Biochemistry, 16*(3), 511–613.

23. Khoja, A. H., Ali, E., Zafar, K., Ansari, A. A., Nawar, A., & Qayyum, M. (2015). Comparative study of bioethanol production from sugarcane molasses by using *Zymomonas mobilis* and *Saccharomyces cerevisiae*. *African Journal of Biotechnology, 14*(31), 2455–2462.
24. Lee, J. M., Pollard, J. F., & Coulman, G. A. (1983). Ethanol fermentation with cell recycling: Computer simulation. *Biotechnology and Bioengineering, 25*(2), 497–511.
25. Moreno, A. D., Ibarra, D., Fernández, J. L., & Ballesteros, M. (2012). Different laccase detoxification strategies for ethanol production from lignocellulosic biomass by the thermotolerant yeast *Kluyveromyces marxianus* CECT 10875. *Bioresource Technology, 106*, 101–109.
26. Nahvi, I., Emtiazi, G., & Alkabi, L. (2002). Isolation of a flocculating *Saccharomyces cerevisiae* and investigation of its performance in the fermentation of beet molasses to ethanol. *Biomass and Bioenergy, 23*(6), 481–486.
27. Ngwenya, T. T., Shukla, P., Baboolal, N., Permaul, K., & Singh, S. (2012). An industrial perspective of factors affecting molasses fermentation by *Saccharomyces cerevisiae*. *Journal of Brewing and Distilling, 3*(2), 23–28.
28. Palmonari, A., Cavallini, D., Sniffen, C. J., Fernandes, L., Holder, P., Fagioli, L., & Mammi, L. (2020). Characterization of molasses chemical composition. *Journal of Dairy Science, 103*(7), 6244–6249.
29. Panesar, P. S., Marwaha, S. S., & Kennedy, J. F. (2006). *Zymomonas mobilis*: An alternative ethanol producer. *Journal of Chemical Technology & Biotechnology, 81*(4), 623–635.
30. Park, S. C., & Baratti, J. C. (1992). Continuous ethanol production from sugar beet molasses using an osmotolerant mutant strain of *Zymomonas mobilis*. *Journal of Fermentation and Bioengineering, 73*(1), 16–21.
31. Pimpakan, P., Yongmanitchai, W., & Limtong, S. (2012). Bioethanol production from sugar cane syrup by thermo-tolerant yeast, *Kluyveromyces marxianus* DMKU3-1042, using fed-batch and repeated-batch fermentation in a nonsterile system. *Bioresource Technology, 46*, 582–591.
32. Rasmey, A. H. M., Hassan, H. H., Abdulwahid, O. A., & Aboseidah, A. A. (2018). Enhancing bioethanol production from sugarcane molasses by *Saccharomyces cerevisiae* Y17. *Egyptian Journal of Botany, 58*(3), 547–561.
33. Saini, P., Beniwal, A., Kokkiligadda, A., & Vij, S. (2017). Evolutionary adaptation of *Kluyveromyces marxianus* strain for efficient conversion of whey lactose to bioethanol. *Process Biochemistry, 62*, 69–79.

34. Silbir, S., Dagbagli, S., Yegin, S., Baysal, T., & Goksungur, Y. (2014). Levan production by *Zymomonas mobilis* in batch and continuous fermentation systems. *Carbohydrate Polymers, 99*, 454–461.
35. Singh, G., & Jain, P. L. (1994). Multifractal analysis of emulsion collisions at 10.6A GeV. *Physical Review C, 50*(5), 2508–2516.
36. Sousa, C. C. D., Gonçalves, G. T. I., & Falleiros, L. N. S. S. (2018). Ethanol production using agroindustrial residues as fermentation substrates by *Kluyveromyces marxianus*. *Industrial Biotechnology, 14*(6), 308–314.
37. Thani, A., Laopaiboon, P., & Laopaiboon, L. (2017). Improvement of a continuous ethanol fermentation from sweet sorghum stem juice using a cell recycling system. *Journal of Biotechnology, 251*, 21–29.
38. Wu, W. H., Hung, W. C., Lo, K. Y., Chen, Y. H., Wan, H. P., & Cheng, K. C. (2016). Bioethanol production from taro waste using thermo-tolerant yeast *Kluyveromyces marxianus* K21. *Bioresource Technology, 201*, 27–32.
39. Yuan, W., Zhao, X., Chen, L., & Bai, F. (2013). Improved ethanol production in Jerusalem artichoke tubers by overexpression of inulinase gene in *Kluyveromyces marxianus*. *Bioresource Technology, 18*, 721–727.
40. Zafar, S., & Owais, M. (2006). Ethanol production from crude whey by *Kluyveromyces marxianus*. *Biochemical Engineering Journal, 27*(3), 295–298.

CHAPTER 5

THE ROLE OF FERMENTATION REACTIONS IN CHANGING THE COLOR OF FOODS

LEARNING OBJECTIVES

At the end of this chapter, readers will be able to:

1. Understand color changes during the fermentation of various food products.
2. Understand the factors influencing color changes in fermented meat and fish products.
3. Analyze the impact of processing conditions and ingredients on color in functional meat products.
4. Understand mechanisms of color changes in fermented dairy products.
5. Know the factors influencing milk and cheese color.
6. Recognize yogurt fermentation's effects on color and pH.
7. Understand the processing and color preservation in butter and fermented vegetables.

INTRODUCTORY EXAMPLE

Artisan Meats Inc., a revered name in the realm of gourmet dry-cured meats, prides itself on time-honored techniques to craft exquisite flavors and textures. As they continually strive for excellence, the company recently embarked on a quest to deepen their understanding of color transformation in their products. Aware of the pivotal role fermentation plays in this process, they are determined to optimize their methods to both enhance color stability and preserve the authentic essence of their offerings.

In their pursuit of perfection, Artisan Meats Inc. has set their sights on refining the visual appeal of their dry-cured meats without compromising on taste. Through meticulous research and experimentation, they aim to uncover the nuances of fermentation's impact on color changes. By delving into the science behind these transformations, the company seeks not only to elevate the aesthetic allure of their products but also to maintain their commitment to crafting premium meats that captivate both the eyes and the palate.

UNIT INTRODUCTION

The color of a food plays a crucial role in determining its quality. Sometimes a product can be rejected solely based on its color, even before considering other important factors like texture, aroma, and taste. Color and appearance, along with overall quality, are important factors in the visual perception of all foods. There are three distinct categories to consider when examining appearance: cesia, color, and spatial properties or spatiality (Sharma et al., 2020).

The optical power spectrum characteristics of the stimulus that viewers perceive are associated with color. Translucence, gloss, sheen, haze, lightness, opacity, and matt are examples of cesia, which are attributes of food that can be measured by human observation in terms of their ability to reflect, transmit, or diffuse light. There are two primary categories for spatial features:

- Appearance modes where materials with metallic, pearlescent, or iridescent properties change color based on the observation angle concerning the light incident angle.
- The appearance of surfaces or objects is determined by their optical properties, which are influenced by various factors like textures, patterns, and finishing features. These factors, such as roughness and polish, playing a role in how food looks.

There are two main methods for analyzing food color: chemical analysis, which involves studying the pigments present, and physical analysis, which involves measurement about how light interacts with the food. The final option is backed by objective values, which are formed by various combinations of tristimulus values like CIE XYZ, Hunter Lab, or CIELAB. Using the CIELAB color space as an internal standard, with D65 as the illuminant and 10° as the standard observer, has been widely adopted (Wang & Fung, 1996).

L* represents the degree of lightness, with 0 indicating black and 100 indicating white. When it comes to color, a* value can tell us about redness, with high values indicating a more intense red. On the other hand, large negative values of a* can indicate greenness. As for b* values, they give us insights into the yellowness or blueness of a sample (AMSA, 1991). The color saturation is represented by C* and the hue is denoted as hab. Reflectance measurements can be utilized as an alternative approach for determining color. This method is highly effective for analyzing the quantity of pigment and the chemical state of pigments in their original location (Yu et al., 2021).

This method is closely related to how the brain and eyes perceive vision. This method allows for multiple measurements to be taken on the same sample over a while. Furthermore, the process is quick and relatively simple. The reflectance measurements in foods are influenced by various factors such as surface moisture, fat content, food structure, and the presence of pigment and additives concentrations. The light-scattering properties are closely linked to the structures of food. Reflectance spectra can be used to describe the colors of various foods. This chapter offers a thorough analysis of the color changes that arise with a variety of food items throughout the fermentation process, including dairy products, seafood, meat, and vegetables like olives (Rhee et al., 2011).

5.1. COLOR CHANGES IN DRY-CURED MEAT PRODUCTS

Learning Objectives:

- Explore the influence of microorganisms on color changes in dry-cured fermented meat products.
- Analyze the effects of processing conditions on color stability in dry-cured fermented sausages.

5.1.1. Fermented Meat Products

Microorganisms play a crucial role in the color and other characteristics of dry-cured fermented meat products, making them highly significant among meat products. There are two categories of products in this field: whole-muscle dry-cured meat products and dry-cured fermented sausages. The former includes dry-cured hams like Teruel ham, Serrano ham, Parma ham, Guijuelo Iberian ham, San Daniele ham, and more. The other category refers to dry-cured fermented sausages (DCFS). Many dry-cured meat products have origins in Mediterranean countries and have since become popular worldwide (Kumar et al., 2017).

Dry-cured fermented sausages aretypes of food that falls into the category of intermediate-moisture foods. These sausages have the unique ability to be stored at room temperature without the risk of spoiling. The vibrant color of these sausages is the result of a chemical reaction between certain compounds and myoglobin, a protein found in meat. This reaction leads to the formation of a bright red pigment called nitrosyl myoglobin.

The vibrant red color is a crucial quality characteristic. If this attribute is flawed, it can negatively impact the consumer's purchasing decision. It is worth noting that the absence of what is commonly perceived as a "normal" color can lead to a negative perception of acceptability and a lack of trust in terms of safety (Wang et al., 2021).

From a macroscopic perspective, the DCFS color exhibits non-homogeneity. These features are dependent upon several attributes such as:

- The muscular tissues in dry-cured products have superior integrity compared to fine-paste products.
- The visual characteristics of the meat and nonmeat elements and the physical characteristics of the meat product, such as the composition of the muscle and fat fractions.
- Spaces within and between muscles.
- Formation of emulsions.
- The inclusion of air within the structural matrix enables the microbiota to establish itself and initiate its metabolic processes (Stadnik & Kęska, 2015).

These properties have a direct impact on the color and the movement of water towards the outside. It is important keeping in mind that during the drying process, the product's structure may undergo deformations due to the tissues, water, and cell movements, particularly affecting the cell membranes. Recently the incorporation of starter cultures in industrial DCFS has gained significant recognition as a method to enhance the rate of processing and ensure consistent product quality.

The microbiota in these starter cultures is chosen based on uses of nitrate, nitrite, or a combination of both as curing agents. Thus, the starter cultures used in the production of nitrite-added fermented sausages are composed exclusively of staphylococci and lactic acid bacteria (LAB). On the other hand, starters for nitrate-added sausages include staphylococci, lactic acid bacteria, and micrococci. These bacteria exhibit nitrate reductase activity (Talon et al., 2007).

Did You Know?

Dry-cured meats like prosciutto and salami develop their rich red color through a process called nitrate-to-nitrite conversion during fermentation, which helps maintain their vibrant appearance.

In both industrial processing and traditional (spontaneous fermentation), the microorganisms naturally present or intentionally introduced primarily break down the carbohydrates, resulting in a decrease in the sausage's pH value. As a result, the muscle protein coagulates, leading to a lighter color and a product that is firm, sliceable, and holds together well.

Using starter cultures have the benefit of minimizing variations between commercial batches, due to the microbiota. It is important to consider that starter cultures have multiple functions. They can inhibit the growth of harmful bacteria and prevent spoilage. Additionally, they can improve the flavor and taste of the product, reduce the levels of nitrite or nitrate, and enhance its color properties.

From a chemical perspective, the vibrant red color is a result of the presence of nitrosyl myoglobin (MbFeIINO). The formation of this chemical occurs when nitric oxide (NO) binds to the central FeII atom in the heme group of myoglobin. The color development is influenced by the conversion of nitrite into NO through reduction. Furthermore, the participation of the NO radical as a potential intermediary in the bacterial conversion of nitrite to dinitrogen emphasizes the importance of nitrogen dissimilation in meat-starting cultures within the meat matrix (Papavergou et al., 2012).

Staphylococci might be potentially generate nitric oxide (NO) by using nitrate/nitrite as an alternate electron acceptor in their respiratory chain. *Staphylococcus* carnosus is considered to convert nitrate into nitrite by a nitrate reductase enzyme is located in the cell membrane and plays a role in conserving respiratory energy.

According to Gøtterup and colleagues, the activities of nitrite reductase and nitrate in starter cultures of meat have long been believed to play a crucial role in the color development of fermented cured meat products. The reduction of curing agents into NO is the crucial step in the formation of MbFeIINO.

Numerous studies have been conducted on the utilization of particular starter cultures for the production of fermented meat products. According to Hugas, *lactobacilli 254* strains were identified from naturally fermented dairy products in Spain. The authors observe that *Lactobacillus's* sake was the major strain, accounting for 55% of the isolates. *Lactobacillus curvatus* followed at 26%, while *Lactobacillus bavaricus* and *Lactobacillus plantarum* accounted for 11% and 8% of the isolates, respectively (Singh et al., 2012).

Currently, the most commonly utilized strains in these meat products are *L. sake* and *L. curvatus*. Researchers have found the use of a specific strain of starter culture called *L. casei* subsp. casei-1.001 has been found to greatly enhance the red color in traditional dry Chinese sausages.

Typically, when the pH of meat decreases due to the metabolism of carbohydrates by microbiota, the color of the meat tends to become lighter. This is due to the meat's water-holding capacity (WHC) decreases, leading to a rise in the light-scattering properties within the muscle fiber.

Water is a crucial component in the color of DCFS. Nevertheless, there is no simple model that adequately explains the correlation between water and its properties, including reflection, or the constituents of color, such as the yellow-blue component or red-green component. This is due to the fact that water possesses exceptional characteristics, including a high boiling point, latent heat of evaporation, high dielectric constant, and low density among other solvents (Barbosa et al., 2009).

The strong molecular interactions of these qualities are responsible for some properties of the DCFS. This is due to the correlation between water and the diverse constituents of these items. The reflectance spectra of various dry-cured fermented sausages are shown in Figure 5.1. The metabolic activity of microorganisms can modify the condition (free, semi-bound, and bound) and availability of water, hence impacting the color of DCFS.

The color characteristics of various meat products are regulated by the composition, microbiota type, curing chemicals, metabolic activity, and processing conditions, which cause chemical and biochemical changes.

Fat plays a significant influence in coarse-ground dry-cured fermented sausages. According to Pérez-Alvarez, the presence of fat can impact the color coordinates by causing an increase in fat content, lightness (L*), and yellowness (b*) values, while decreasing the redness (a*) value. Pérez explained the behavior of color coordinates during the procedure (Flores & Toldra, 2011).

In dry-cured meat products, the L* value tends to rise during fermentation and then decline during ripening due to dehydration process, which reduces the amount of free water on the surface. The intensity of redness is heightened due to the formation of nitroso myoglobin and the removal of moisture, which lessens the dilution effect of water in the meat product. In addition, the b* value is influenced by changes in myoglobin states and alterations in the DCFS microstructure that occur during processing (Todorov et al., 2017).

As previously mentioned, fat plays a crucial role in determining the color of DCFS. It was found that sausages with a fat content ranging from 10% to 20% and subjected to high ripening temperatures (24°C–26°C) exhibited favorable color development. Additionally, these sausages showed rapid declines in water activity, pH, and lactic acid production by LAB, highlighting their positive technological characteristics. According to the authors, when fat concentration increases under the same conditions, it leads to a lack of color development (Little, 1998).

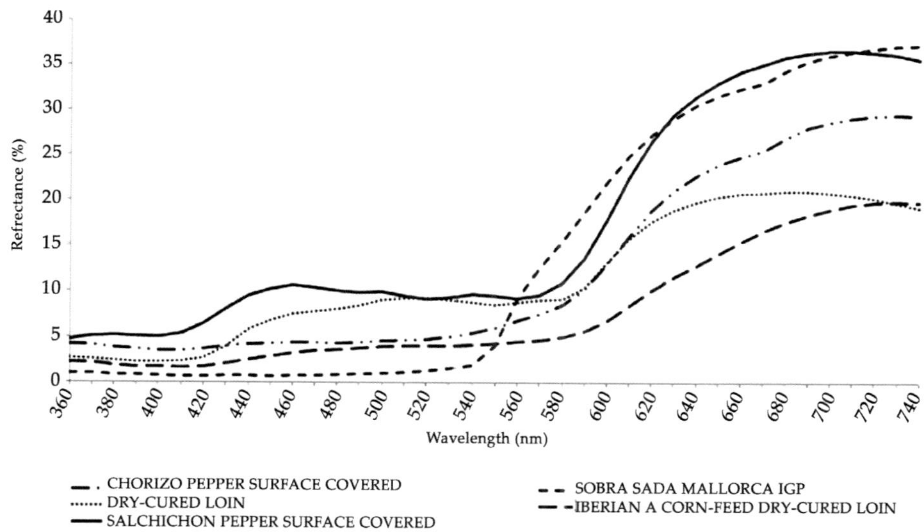

Figure 5.1. *Illustration of reflectance spectra of various dry-cured fermented meat products.*

Source: Esther Sendra, Creative Commons License.

5.1.2. Color Stability

There are factors that can impact the stability of color in this particular type of meat product. Various factors can influence the fermentation reaction, such as temperature, pH, relative humidity, lipid oxidation, microbiota load and type, partial oxygen pressure, type of muscle, metmyoglobin-reducing systems, light intensity, and antioxidant concentration among all. (Suman et al., 2014).

Many lactobacilli can produce hydrogen peroxide, which can cause the nitroso heme pigment to change color. Certain microbial strains with catalase activity can be used in conjunction with catalase-positive cocci to overcome above issue. The fact that meat and meat products packaged in a protective environment are so popular shows how important it is to stop or delay meat discoloration (Falkensammer et al., 2013).

5.1.3. Processing Conditions

Temperature is playing a crucial role in influencing the behavior of microbiota in DCFS as a processing parameter. This parameter is closely linked to fermentation condition, ripening, and drying. Therefore, temperature changes can impact various factors such as pH, water activity, texture, microbial growth and metabolism, and color (Jansen et al., 1998).

DCFS's unique cured color develops during the first processing step (mixing, resting), and several scientific and technological aspects affect its long-term durability. Therefore, sausages with a larger casing diameter (>80 mm) necessitate the implementation of fermentation conditions at 10°C–12°C (slow fermentation). These temperatures help to reduce the high metabolic activity of microorganisms (Jiménez, 2007).

When the temperatures are adjusted, the lactic acid formation decreases, preventing the sausages from developing a strong acid flavor and lighter or less reddish color. During the ripening process, Soyer observed that the formation of nitrosyl myoglobin occurs rapidly when the temperature is increased. Therefore, the appealing red hue of DCFS is enhanced and evenly distributed (Dennis et al., 2001).

During the process of ripening, it is important to consider the formation of mold on the surface. Molds have two primary functions in terms of color. They utilize lactic acid as a substrate, hence raising the pH of DCFS. By increasing the water holding capacity (WHC) and altering the optical properties (specifically reducing light-scattering qualities), the desired effects are achieved. Additionally, these alterations inhibit the diffusion of oxygen through the sausage, hence minimizing oxidation reactions that can lead to the fading of the dry-cured color (Chinnaswamy & Bhattacharya, 1983).

The process of technology leaves distinct markers, and the spectrum can be utilized to measure the extent to which the color of DCFS is modified by these processes. Furthermore, apart from reflectance spectra, it is necessary to take into account other factors in the spectrum, including the characteristics of the surface (smooth or rough), the presence of nonmeat substances in the particles, and the dispersion or dissolution of molecules in the aqueous phase (such as salt or phosphates) (Mosqueda et al., 2004).

5.1.4. Functional Fermented Meats

A basic adaptation of Dipplock's definition of a functional meat product can be made for DCFS, even though there is no precise definition for this term. Therefore, a functional DCFS has to remain the same and show its effects in levels that are often expected to be taken as part of a regular eating pattern. The problem facing the meat business is in the production of aesthetically pleasing functional meat products that satisfy consumer needs for affordability, quality, and health while maintaining the traditional flavor and color of the meat (Munekata et al., 2022).

The consideration of color holds significant importance in the development of fresh meat products with enhanced functionality. It is important to considervarious factors, including biochemical, chemical, microbiological, and physical, are closely interconnected when producing a functional DCFS. Therefore, when a factor is altered, the color characteristics and stability may vary significantly from the expected result. In functional meat products, various ingredients like probiotics, olive oil, polyphenols, and dietary fiber are commonly incorporated into the formula (Chaijan et al., 2005).

These ingredients have the potential to change the metabolic patterns of various microorganisms, resulting in potential modifications to the color. Probiotic meat-starting cultures seem to be a suitable option for inclusion in DCFS. However, it is important to ensure that the addition of this microbiota does not compromise the technological and sensory characteristics of the product, as this could potentially result in its rejection (Arihara, 2006).

Many researchers have suggested incorporating probiotics into the production of DCFS, but several challenges need to be addressed to achieve success. There are several significant challenges to consider when it comes to produce DCFS meat products. These include the high content of curing salt, the meat environment, and the low pH and water activity that occurs during drying and fermentation. Additionally, the impact of various functional ingredients on the color of these products has been studied (Leroy et al., 2006).

When analyzing food, food technologists need to examine the impact of functional ingredients on various factors such as lipid oxidation, protein-lipid interaction, and modifications in microbiota, pH, and water-holding capacity. This helps in effectively managing the impact of these new components (Kołożyn & Dolatowski, 2009).

5.2. COLOR CHANGES IN FERMENTED FISH PRODUCTS

Learning Objectives:

- Understand the factors influencing the color of fresh fish muscle.
- Explore the role of myoglobin, carotenoids, and other pigments in determining fish muscle color.
- Understand the color changes in fermented fish products and their relation to processing conditions.

5.2.1. The Color of Fresh Fish

The color of fish, whether fresh or processed, plays a crucial role in influencing consumers' purchasing decisions. The uniformity of the color is particularly significant as it serves as an important quality parameter. Myoglobin plays significant role in determining the color of muscle, which is influenced by its redox states (red oxymyoglobin [OxyMb], purple reduced myoglobin [DeoMb], and brown metmyoglobin [MetMb]) as well as its concentration (Farah et al., 2011).

In addition, there is a strong correlation between the alterations in myoglobin states and lipid oxidation. The oxidation of myoglobin promotes lipid oxidation in fish muscle, resulting in changes to the muscle. Myoglobin consists of a single polypeptide chain called globin, along with a prosthetic heme group. This heme group is a complex of iron (II) protoporphyrin-IX, with a molecular weight ranging from 14–18 kDa. It is worth noting that fish myoglobin tends to be smaller in size compared to mammalian counterpart (Ríos et al., 2022).

The stability of myoglobin differs among species, as a result of variations in amino acid sequences and secondary structures of globin. Chaijan's research findings revealed that myoglobin from sardines (Sardina pilchardus) is susceptible to oxidation and denaturation when exposed to temperatures above 40°C and extremely acidic or alkaline pH levels. The rate of autoxidation of myoglobin is directly related to its stability, which in turn affects the color and lipid oxidation of the meat. Tuna is renowned for its exceptional red muscles, which contain high concentrations of myoglobin (Olgunoglu, 2013).

Fun fact:

In traditional Japanese cuisine, fermented fish products like "narezushi" and "shiokara" undergo a unique fermentation process that not only enhances flavor but also results in distinct color changes, adding visual interest to the dish.

The color of fish muscle is also influenced by the presence of several natural pigments, including rainbow trout or melamine in crustacea and carotenoids in salmon, which contribute to the retention of color in the muscle. The red-to-pink muscle color of rainbow trout is a crucial factor in determining its quality. This color is attributed to astaxanthin, a dietary carotenoid (3,3¢-dihydroxy-β,β-carotene-4,4¢-dione) (Tavares et al., 2021).

In his research, Bartel made an interesting discovery about the relationship between fish muscle color and carotenoids. He found that the total amount of carotenoids and the ratio of red pigments (canthaxanthin and astaxanthin) to yellow pigments (zeaxanthin and lutein), played a significant role in determining the color. Bartel's research revealed that rainbow trout, known for its red-to-pink hue, contained 0.212 g/g of carotenoids in terms of fresh weight (Lu et al., 2010). The primary sources of these carotenoids were astaxanthin and canthaxanthin. Conversely, the number of carotenoids in the white muscle of sardines was determined to be 10.537 μg/g of fresh weight.

The variations in myoglobin content, lipid content, different myoglobin states, different carotenoids, and the development of lipid oxidation all play a role in determining the color of fish muscle. This can be assessed conveniently using the CIELAB color space (Gelman et al., 2001).

The authors of this study discussed the use of a* and b* coordinates to classify the color of fish muscle. Red and yellow fish muscle is typically characterized by higher a* and b* values, while white fish muscle tends to have lower a* and b* values (Zhai et al., 2017). The CIELAB color coordinates for the lightness (L*), redness (a*), and yellowness (b*) of various fish species are displayed in Table 5.1.

5.2.2. The Color in Fermented Fish Products

Processed fish products and fish play a crucial role in global food consumption. Fish fermentation is a traditional method of preserving fish, which can be applied to both marine and freshwater fish. This preservation process improves the sensory and sanitary characteristics of the product. Most fish fermentation still occurs at the home or small-scale level by spontaneous mechanisms in some areas.

The initiation of a spontaneous fermentation process can be quite time-consuming, resulting in products of varying quality. This variability can limit their acceptance and commercial significance. Leistner identified a total of 58 different types of lactobacilli and cyanobacteria in gravid salmon (Panda et al., 2011).

Table 5.1. *CIELAB Color Coordinates of Different Fish Species*

	a*	b*	L*
White Tuna (*Sarda sarda*)	1.3	0.9	31.8
Hake (*Merluccius merluccius*)	–0.1	–2.6	48.0
Sardine (*Sardina pilchardus*)	8.2	2.1	37.2
Sea Bass (*Dicentrarchus labrax*)	1.5	–1.2	40.7
Rainbow Trout (*Oncorhynchus mykiss*)	10.3	9.7	37.2
Tuna (*Thunnus thynnus*)	17.2	5.8	31.6
Salmon (*Salmo salar*)	11.8	13.4	48.3

Source: Maria E., Creative Commons License.

Extensive research has been conducted on the starter culture utilization in food fermentation, with the aim of enhancing processing efficiency and ensuring consistent product quality in commercial settings. In recent years, there has been a growing interest in the utilization of lactic acid bacteria for fish products (Feng et al., 2021).

Lactic acid fermentation results in the rapid acidification of the raw material by producing organic acids such as lactic acid and acetic acid. Additionally, it generates antimicrobial substances that effectively inhibit the growth of harmful food microorganisms. There are different methods for developing fermentation processes for fish, including working with whole fish fillets or minced fish (Huda, 2012).

5.2.2.1. Noncomminuted Fermented Fish

The conventional fermentation method involves preserving fish fillets by combining sugar and salt (to achieve a final concentration of 10 and 25 g per kilogram of fish, respectively),along with incorporating herbs or spices. The mixture is evenly applied by pressing it over the inner side of the fillets. The fillets are then placed on a metal rack and stored, often at a temperature of 12°C, overnight. This allows the ingredients to spread throughout the meat by diffusion. Afterward, the ingredients that have not dissolved are taken off the surface of the fillets (Sangjindavong et al., 2008).

After that, the fillets are carefully placed in specialized bags andin a controlled environment with temperatures ranging from 12°C to 35°C. This process can last anywhere from a short duration of 72 hours to an extended period of several years. The fermentation characteristics, including time and temperature, vary depending on the type of fish and the location where the fish are processed. For example, in several European countries, fatty fish like herring, salmon, sardine, white tuna, mackerel, and tuna are commonly preserved through spontaneous and salting fermentation over a brief duration. Asia, the Philippines, Japan, Korea, and other countries are known for their long fermentation periods (Fiddler et al., 1984).

The fermented fish products' color is influenced by various process characteristics, including the color of the raw fish fillets. These factors collectively determine the final result. Various authors have noted that the variations in color parameters of the raw material can be reduced through the salting and fermenting process. The CIELAB color coordinates of fermented and salted fish fillets from various species of fish are shown in Table 5.2.

In research conducted by Morzel, the study aimed to observe the variations in color of fermented salmon fillets. The experiment involved subjecting the fillets to low temperatures (12°C) and a short fermentation period (72 hours). The study discovered that the color coordinates in the salmon fillets experienced a decrease during overnight storage at 12°C, followed by a gradual increase throughout the remaining fermentation process. The a* and b* coordinates experienced decay of 3–8 and 4–10 units, respectively, suggesting a reduction in redness and yellowness when compared to fresh salmon. This change can be linked to the salting process (Morzel et al., 1997).

In comparison to initial time, the fillets regained their red and yellow color during the fermentation and storage period. During the first processing step (overnight storage), the L* values dropped by 8–10 units, suggesting that samples became darker in comparison to fresh salmon—likely as a result of salting. At 12°C, the fillets' brightness is produced in a fairly comparable way during fermentation (Li et al., 2023).

Table 5.2. *CIELAB Color Coordinates of Fermented and Salted Fillets from Different Fish Species*

	a*	**b***	**L***
White Tuna (*Sarda sarda*)	1.7	6.2	43.5
Tuna (*Thunnus thynnus*)	8.9	5.8	37.8
Salmon (*Salmo salar*)	11.6	12.4	47.3
Sardine (*Sardina pilchardus*)	2.3	6.6	45.5

Source: Esther Sendra, Creative Commons License.

5.2.2.2. Fermented Minced Fish

Fish sauce and fish paste are most popular products, known for their savory taste and distinctive scent that can stimulate the appetite. Fish paste is made by naturally fermenting whole or minced fish or shrimp and adding 20–25% salt while the mixture is left at room temperature to ferment under ambient conditions. Minced fish that has undergone fermentation is a popular protein-rich food in Southeast Asia.

Typically, it consists of spices, salt, a source of carbohydrates, and freshwater fish. Fish fermentation can be started with a starting culture,or it can happen naturally. According to research, using a minced fish, salt, and glucose system with starting cultures of *Lactobacillus* plantarum and *Pediococcus pentosaceus* speeds up the fermentation process at 30°C (Gao et al., 2019).

In a study conducted by Yin, and Pan, various lactic acid bacteria (LAB) were utilized as starter cultures to ferment mackerel mince at a temperature of 37°C. The LAB used included *Lactococcus lactis*, *L. plantarum*, and *L. helveticus*. Similarly, Glatman conducted an experiment where *L. plantarum*, *P. pentosaceus*, and *Leuconostoc mesenteroides* were injected into minced frozen fillets of yellowfin tuna and fermented at a temperature of 8°C. The product's sensory characteristics, including color, and its ease of digestion have been improved in every case (Riebroy et al., 2008).

However, fish sauce is a clear liquid with a straw yellow to amber color. It is produced by fully hydrolyzing a mixture of fish and salt for 9–12 months. In Korea, salted and fermented anchovy sauce is a commonly used ingredient to enhance the distinct flavors of many traditional dishes. The sauce is often referred to as having an "amber" color (Neiva et al., 2011).

5.3. COLOR CHANGES IN FERMENTED DAIRY PRODUCTS

Learning Objectives:

- Understand the mechanisms influencing color changes in fermented dairy products.
- Explore the impact of milk composition, processing techniques, and microbial metabolism on dairy product color.

When it comes to dairy meals, people initially notice the color, which can also influence their perception of other qualities like aroma and flavor. There are three main categories of fermented dairy products: cheese, yogurt other fermented milks, and butter. This section will examine the mechanisms responsible for color changes that occur throughout the production and storage of these products. These mechanisms mostly include alterations in protein structure due to changes in pH, dehydration, coagulation, and microbial metabolism. Coloradditions will be excluded from this discussion (Milovanovic et al., 2020).

Riboflavin, found in milk whey, gives a yellow-greenish hue, while carotenoids in milk fat contribute to its yellowish color. The primary cause of the white appearance of milk is the dispersion of visible light by casein micelles and fat globules. Caseins exhibit superior dispersion of blue light compared to red light, resulting in skimmed milk possessing a distinctive bluish hue. Various technological processes alter the color of milk.

The application of heat to milk results in an increase of its white hue (for reasons that are currently unknown), and more intense heat treatments lead to browning as a result of Maillard reactions. It is proposed that the cause of the milk whitening phenomenon in low-heat milk could be attributed to either the presence of denatured whey proteins, resulting in an increase in dispersed particles, or a rise in colloidal phosphate in casein micelles (Hernández et al., 2022).

The fat globules are broken down into smaller sizes, resulting in milk that appears whiter by the process of milk homogenization. This is because the smaller fat globules scatter light more effectively, giving the milk a brighter appearance. Therefore, the color of milk and its riboflavin and carotenoid concentration will change depending on factors that affect fat globules or the structure and integrity of casein.

One of the key factors that influence the color of milk is its origin. In comparison to milk from other milking ruminants, cow's milk tends to have a yellower hue. This can be attributed to the transfer of carotenoids to the milk, particularly during the late spring and summer seasons when cows are grazing on certain types of feed. Research has shown that the diet of cows can have an impact on the color of their milk. Specifically, cows are fed a diet primarily consisting of grass tend to produce milk with a more vibrant yellow color compared to cows are fed hay or concentrates. The carotene content in milk can be influenced by the genetics of the specific breed (Rankin & Brewer, 1998).

Goats and sheep do not transfer carotenoids to their milk, resulting in a whiter appearance compared to cow's milk. The variation in color among different types of milk has sparked interest in the process of lightening cow's milk for the production of cheeses that are highly sought after for their whitish appearance, like mozzarella and white-brined cheeses. Carotenoids can be bleached by utilizing peroxides such as H_2O_2 or benzoyl peroxide, or they can be masked with substances like chlorophyll or titanium oxide (Sharma et al., 2023).

5.3.1. Cheese

Multiple authors have examined the mechanics involved in the process of cheese production. Essentially, the process of cheese manufacture can be broken down into five key steps: coagulation, acidification, syneresis (the removal of whey), molding, and salting. The phases have some overlap, and modifications in the conditions of each step allow the cheese maker to regulate the composition of the cheese, so influencing its ripening process and overall quality. The primary components of milk proteins consist of caseins, which made by up approximately 70% to 80% of the total, and whey proteins, which account for about 20% to 30%. Caseins play a crucial role in the formation of the curd matrix. They are composed of four types of phosphoproteins: $\alpha s1$, $\alpha s2$, β, and κ.

Cheese made from ultrafiltered milk or milk that has undergone a thorough heat treatment is the only source of whey proteins. Under such circumstances, the whey proteins bind to the surface of caseins. Phosphorylated residues in caseins aid in the binding of calcium ions and contributing the acidic characteristics of caseins, with an isoelectric point around 4.5 to 5.0 (Beresford et al., 2001).

Caseins, together with calcium phosphate and calcium, form aggregates consists of many casein micelles. The α and β caseins are mainly situated on the inside, whereas the κ caseins, being glycoproteins, are primarily located on the exterior. This arrangement helps protect the α and β caseins from any unintended precipitation or aggregation (Everett & Auty, 2008).

Did You Know?

The color of cheese can vary widely depending on factors such as the type of milk used, the presence of additives like annatto or carotene, and the duration of fermentation. For example, aged cheeses like cheddar develop deeper yellow hues due to prolonged fermentation.

Milk coagulation is often accomplished by either enzymatic activity or acidity. The majority of cheeses are produced using enzymatic coagulation, a process that includes altering casein micelles through controlled proteolysis. Coagulation enzymes catalyze the hydrolysis of κ casein at the Phe105-Met106 bond, results in the exposure of the α and β caseins. When calcium ions are present, the casein micelles come together to create a curd, which traps fat globules and other components of milk. Enzymatic coagulation consists of two distinct phases: The initial phase starts with the inclusion of coagulant enzymes until approximately 85% of κ-casein is hydrolyzed. The subsequent phase begins from 85% hydrolysis onwards, with the crucial involvement of ionic calcium (Farkye, 2004).

The color changes resulting from enzymatic coagulation are shown in Figure 5.2. The experiments were conducted using commercially pasteurized milk, and the color was evaluated every 5 minutes until coagulation was finished within 45 minutes. The primary alterations occurred during the initial 20 minutes, likely corresponding to the initial coagulation phase, and reached a stable state. It is worth mentioning that partially skimmed and skimmed milk showed more pronounced color changes in comparision with whole milk, consistently moving towards the color values of whole milk (McSweeney et al., 2004).

During the fermentation process, coagulant enzymes work with starter cultures to induce acidification. The process of acidification may take place in two ways. In the homo fermentative route, over 90% of the lactose is converted into lactic acid. On the other hand, the heterofermentative route also produces lactic acid from lactose hydrolysis, but it leads to the formation of other compounds. The acidification process has various effects, including the alteration of coagulant activity, the solubilization of colloidal calcium phosphate, the promotion of syneresis, and the influence on enzyme activity during ripening (Lawrence et al., 1984).

When milk coagulation occurs mainly due to acidification, it is usually caused by the fermentation of lactose by lactic acid bacteria. Destabilization of caseins occurs as the pH falls and approaches their isoelectric point; this results in the formation of a flexible gel in which caseins are not harmed. Observing variations in color resulting from the proliferation of lactic acid bacteria. The different technological factors that affect water retention in the curd play a crucial role in determining the color of cheese. Various factors need to be considered, such as the conditions for cutting and whey removal, the mechanisms of coagulation, and the process of salting, cooking, and ripening. They primarily affect the cheese's color by causing ultra-structural modifications (Bachmann, 2001).

Certain cheeses are consistently made with inclusion of colorants, such as annatto, which primarily determine the appearance of the final product. Some cheeses, like pickled cheese, have a whitish color due to being stored in brine. However, in certain types of cheeses, microbial metabolism plays a crucial role in creating the distinct color. This is observed in cheeses like surface-ripened cheeses, blue-veined cheese and mozzarella.

Mozzarella cheese is often used in baked dishes such as pizza, where the remaining galactose can potentially alter the melted cheese color. Galactose, a type of sugar, can participate in Maillard reactions when combined with an amine, such as an amino acid. This reaction can result in the formation of an unappealing brown color. Browning in mozzarella may be minimized by using galactose-positive cultures of *Streptococcus thermophilus* and *Lactobacillus delbrueckii subsp. Bulgaricus* (Hayaloglu, 2016) (Figure 5.3).

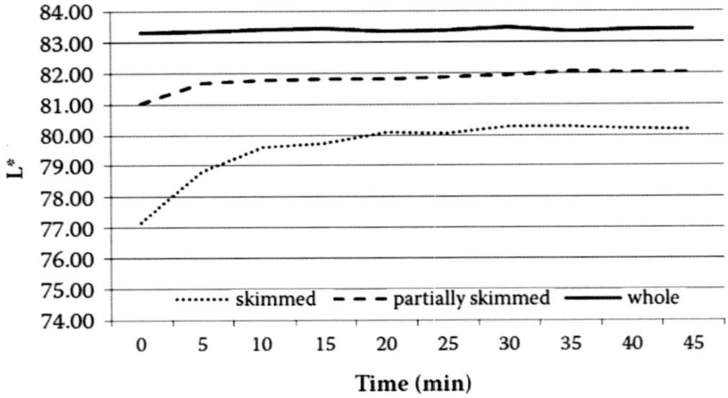

Figure 5.2. *Illustration of assessment of variation of color during milk enzymatic milk coagulation with L*.*

Source: Sayas-Barberá, Creative Commons License.

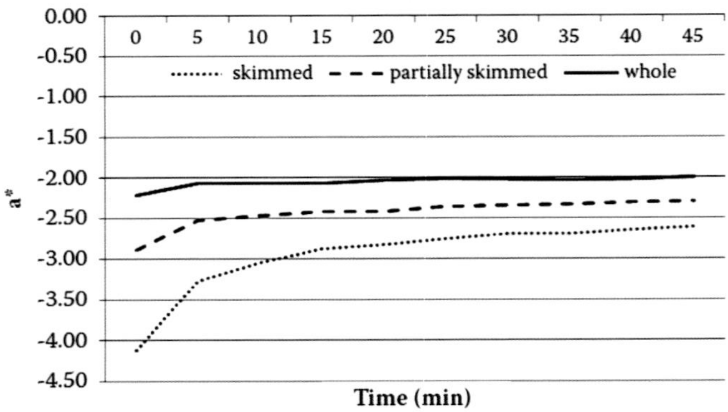

Figure 5.3. *Illustration of assessment of color variations during milk enzymatic milk coagulation with a*.*

Source: Sayas-Barberá, Creative Commons License.

It is important to highlight the issue with processed cheese, as its propensity to turn brown during and after the manufacturing process is undesired. The galactose content in cheese, particularly in mozzarella cheese, plays a crucial role in browning. The maturing period of the cheese before processing directly affects the level of browning since a longer aging period leads to a reduction in galactose content and consequently lesser browning. Quickly cooling processed cheese also decreases cheese browning (McMahon et al., 1999). *Brevibacterium linens* is frequently found in smear-ripened cheeses (El Soda,

2000). The development of carotenoids results in the production of colonies that range in color from yellow to deep orange-red. *Brevibacterium linens* primarily produces the aromatic carotenoid 3,3¢-dihydroxy-isorenieratene, also known as φ,φ-carotene-3,3¢-diol.

The minor constituents comprise a monohydroxy molecule and a hydrocarbon known as isorenieratene. The change in color from yellow-orange to pink-purple takes place when colonies of *B. linens* are coated with an alkaline solution. This transition is caused by the presence of the dihydroxyphenyl carotenoid. Carotenoids can be classified into different categories based on their hydroxylation status, including non-hydroxylated, mono-hydroxylated, and di-hydroxylated forms. The specific type of carotenoid produced depends on the strain and kinetic factors (Mistry, 2001) (Figure 5.4).

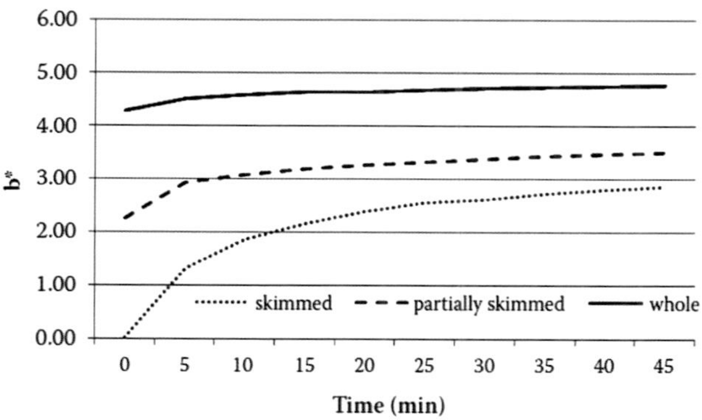

Figure 5.4. *Illustration of assessment of color variations during milk enzymatic milk coagulation with b*.*

Source: Juana Fernández-López, Creative Commons License.

Yeasts, particularly those belonging to the genera *Candida, Debaryomyces, and Torulopsis*, can be found in surface-ripened cheeses. These yeasts have a significant impact on the cheese surface, primarily by raising the pH through the consumption of lactic acid and the production of vitamins such as niacin, riboflavin (which gives color), and pantothenic acid. This, in turn, promotes the growth of micrococci and *B. linens* (Williams, 2005).

The presence of yeast typically diminishes within a period of 1 to 20 days throughout the ripening process, making room for the growth of micrococci and *B. linens*. The growth of yeast is affected by a range of factors, such as the temperature at which it matures. This temperature also has the impact on the color of the cheese's surface. When the ripening temperature is 14°C, the metabolism of yeast is accelerated compared to when it is 10°C. As a result, the growth of bacteria on the cheese's surface is increased. Recently, it has been discovered that other bacteria play a role in the development of smear-ripened cheeses. Research has indicated that the significance of *B. linens* has been overstated, as other bacteria like *Corynebacterium casei, Brevibacterium aurantiacum, Staphylococcus saprophyticus, and Microbacterium gubbeenense*, have been detected in these cheeses as well. These bacteria have been demonstrated to play a role in the coloration of cheese in carefully controlled experiments with cheese curds.

There have been significant advancements in the color of cheeses, particularly in mold-ripened cheeses. These cheeses can be categorized into two primary types: those with white mold called *Penicillium camemberti*, and those with blue veins caused by *Penicillium roqueforti* (Tamine, 2000).

5.3.2. Yogurt

In yogurt making, the primary focus is "the conversion of lactose to lactic acid". This process has been extensively studied and documented in various reference books. Throughout this process, the pH decreases and eventually reaches the isoelectric point of caseins, resulting in alterations to the ultrastructure of milk and a subsequent change in its color. Figures 5.5 and 5.6 show the color and pH variations in whole, half-skimmed and skimmed milk as a result of fermentation by yogurt cultures (Meydani & Ha, 2000).

An experiment was conducted using commercially available UHT milk. UHT milk tends to have higher initial b* values compared to pasteurized milk, which is likely a result of Maillard reactions occurring during the UHT heat processing. During fermentation, the color of milk can go through noticeable transformations due to fluctuations in pH levels. There is a noticeable shift in pH levels at around 5.5, followed by another shift at approximately 5.0. Correlation formulas have been developed to predict the color changes in yogurt as the pH decreases during fermentation (Lee & Lucey, 2010).

The authors of this study found that refrigerating yogurt after fermentation resulted in L*(increased whiteness), b* (increased yellowness), and a* (reduced greenness). No changes were observed in the color components of yogurt during its refrigerated storage for 28 days.

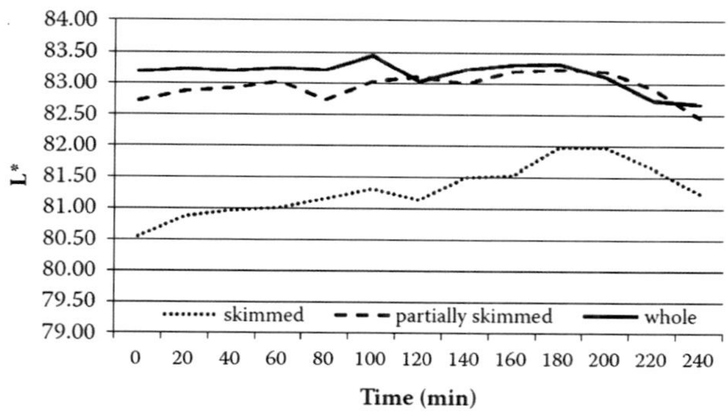

Figure 5.5. *Illustration of assessment of pH and color changes during yogurt fermentation with L*.*

Source: Jose A., Creative Commons License.

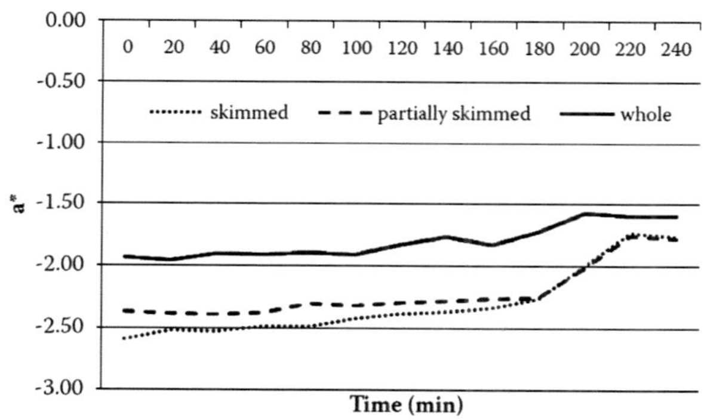

Figure 5.6. Illustration of assessment of color and pH changes during yogurt fermentation with a.*

Source: Pérez-Alvarez, Creative Commons License.

5.3.3. Butter

The butter color mostly relies on the carotenoid's presence inside the milk fat. Butter is typically found in two forms: cultured cream butter and sweet cream butter. When lactic acid bacteria are added to sweet butter, the resulting cultured cream butter has a lower pH and a unique diacetyl flavor (Wright et al., 2001).

Research has indicated that there are no significant variations in butter color as determined by a colorimeter when it comes to the effect of fermentation on butter color. However, when evaluated through sensory analysis, cultured butter is often perceived as having a more vibrant yellow hue and a shinier appearance compareto sweet butter (Mallia et al., 2008).

5.4. COLOR CHANGES IN FERMENTED VEGETABLES EXEMPLIFIED IN OLIVES

Learning Objectives:

- Understand the fermentation process of olives and its impact on color.

Historically, fermenting vegetables has been used to prolong the longevity of perishable vegetables. Vegetables can have their nutritional content, sensory appeal, shelf life, safety improved and preserved by the use of a simple, yet significant biotechnology called lactic acid bacteria fermentation (Rodriguez et al., 2009). Olives are a crucially significant fermented vegetable in the global economy.

When the olive fruit achieves the ideal degree of ripeness, it is harvested from particular kinds of olive trees (Olea europaea sativa). The olive fruit contains a bitter compound known as oleuropein, which is a glucoside. It has a relatively low sugar concentration ranging from 2.6% to 6.0%. Additionally, olives have a high oil content ranging from 12% to 30%, but these percentages can vary depending on the maturity of the fruit and the specific olive variety. These characteristics make it a fruit which can not directly consumpted from the tree (Blatchly et al., 2014).

The olive must go through a variety of processes to become suitable for consumption. These processes can vary significantly from one region to another and are influenced by the specific type of olive being used (IOOC, 2010). The market is dominated by three major industrial preparation types: naturally, black olives, ripe olives by alkaline oxidation, often called the Californian style, and green Spanish style, or Sevillian, which makes up about 60% of production. There are various methods for preparing table olives, which depend on factors such as fermentation conditions (temperature, salt levels, and type of acid) and the type of olives used (green, turning color, or black) (Figure 5.7).

The black or green olives are picked, sorted, and size-graded. The ingredients are placed in a brine solution with a salt content of 6% to 8% NaCl, either in an environment without oxygen or in an environment with oxygen. They undergo a natural fermentation process that lasts for several months. Substrates including carbohydrates (mostly glucose, but also sucrose, mannitol, and fructose) and organic acids (citric, malic, and acetic) are leached as a result of osmotic exchange (Nout, 2001).

Figure 5.7. *Representation of Spanish-style olives commercial preparations.*

Source: Esther Sendra, Creative Commons License.

As a result, the brine solution becomes an ideal environment for the spread of microorganisms. During the processing of green Spanish-style olives, the bitterness is eliminated through a hydrolysis process. This process requires the use of a diluted sodium hydroxide solution for an extended period. The hydro soluble carbohydrates and oleuropein are eliminated and transformed during this period (Ferguous et al., 2023).

The ripe olives processing involves a series of steps where the olives are treated with sodium hydroxide solutions (lyes) for three consecutive days. The olives are suspended in water or diluted brines in between these treatments, with air bubbles moving through them. Ultimately, a solution of lactate and ferrous gluconate is introduced to stabilize the color.

Practice Problem

AgriTon Farm, a leading producer of organic vegetables, is striving to enhance the fermentation process of their pickled olives to improve their visual appeal without compromising on taste. As they seek to understand the mechanisms driving color changes during fermentation, they face the challenge of optimizing their methods to achieve vibrant colors that attract consumers while maintaining the authentic flavor profile of their products. How can AgriTon Farm adjust their fermentation process to achieve desired color outcomes in their pickled olives?

Solutions to Practice Problem

To address this challenge, AgriTon Farm can experiment with various factors influencing the fermentation process. Firstly, they can adjust fermentation duration and temperature to optimize enzymatic reactions and pigment formation. Additionally, by carefully selecting and controlling microbial cultures, they can influence color development during fermentation. Furthermore, AgriTon Farm can explore the use of natural additives or processing techniques to enhance color stability and intensity in their pickled olives. Through systematic research and experimentation, AgriTon Farm can refine their fermentation process to achieve visually appealing colors that complement the exceptional taste of their pickled olives, thus ensuring continued customer satisfaction and brand loyalty.

The fermentation process is driven by the addition of NaCl and gradual decrease in pH favoring the growth of lactic microflora. These microorganisms go from representing only 1% of the overall microbial population in the initial brine to a significant 80% within a short span of time. During natural fermentation, the indigenous lactic acid bacteria (LAB) undergo spontaneous changes. After the process, a significant portion (39.9%) of the total LAB population consists of *Lactobacillus plantarum*. However, other LAB species like *Lactobacillus pentosus* and *Leuconostoc mesenteroides* have also been identified (Majerska et al., 2019).

Gram-negative bacilli are the predominant bacteria during the initial fermentation stage. Their metabolism produces several end products such as carbon dioxide, hydrogen, lactic acid, acetic acid, and ethyl alcohol. In the second phase, due to a decrease in pH (reaching a pH of 4.5), there is a significant development of Lactobacilli, which exclusively creates lactic acid as the end product of glucose fermentation. In the third phase, acid production stops when the pH drops to 4.0 or lower. Once all the sugars are used up, the fermentation process is considered complete, and the storage period begins (Devanthi & Gkatzionis, 2019).

By the fermentation process, olives remain in the brine solution until they have reached the desired level of curing, at which point they are packed and prepared for commercial sale. The olives are processed, graded, and packed according to market demands. During certain commercial presentations, it is possible to break or divide them along their longer axis and enhance their flavor using organic substances. Preservatives or thermal pasteurization are often used in the packaging of fermented vegetables to guarantee their microbial stability throughout storage (Liszkowska & Berlowska, 2021).

SUMMARY

- The chapter explores the important role that fermentation reactions play in changing the color of different food products. The study analyzes the color variations observed in various categories including dry-cured meat, fermented fish, dairy, and vegetables.
- Various factors can influence color, stability in dry-cured meat products including processing conditions, fermentation, and product type. The process of fermentation is essential in bringing about a significant change in the color of fermented fish products, creating a noticeable contrast between the fresh fish color and the color of the fermented variety.
- Furthermore, in fermented dairy products including cheese, yogurt, and butter, the fermentation processes result in distinct alterations in color. The chapter also delves into the color changes observed in fermented vegetables, using olives as a prime example.

REVIEW QUESTIONS

1. How do fermentation reactions affect food color changes?
2. Influence of processing conditions on color stability in fermented meats.
3. Compare color changes in fermented fish and dairy products.
4. Color transformations in olives during fermentation and underlying mechanisms.
5. Impact of pH and chemical changes on fermented food color.
6. Role of enzymes and oxidation in color changes in fermented vegetables.
7. Fermentation's influence on the Maillard reaction in dairy product color.
8. Contrasting effects of traditional vs. modern fermentation on food color.

MULTIPLE CHOICE QUESTIONS

1. What is the primary factor influencing color stability in dry-cured meat products?

a. Salt content

b. Fermentation

c. Processing conditions

d. Packaging

2. Which of the following categories undergoes significant color changes due to fermentation?

a. Fresh fruits

b. Canned vegetables

c. Fermented dairy products

d. Frozen meats

3. In fermented fish products, what aspect is compared regarding color?

a. Freshness

b. Texture

c. Fat content

d. Origin

4. Olives are used as an example to illustrate color changes in which category of food?

a. Fruits

b. Vegetables

c. Grains

d. Legumes

5. What role does fermentation play in altering the color of yogurt?

a. It decreases color intensity.

b. It increases color stability.

c. It contributes to the formation of creamy white to yellowish hues.

d. It has no effect on color.

6. How do processing conditions influence color stability in dry-cured meat products?

a. They have no effect on color stability.

b. They directly affect salt content.

c. They influence the duration and intensity of fermentation.

d. They primarily affect packaging.

Answers to Multiple Questions

1. (c); 2. (c); 3. (a); 4. (b); 5. (c) 6. (c)

REFERENCES

1. Arihara, K. (2006). Strategies for designing novel functional meat products. *Meat Science, 74*(1), 219–229.
2. Bachmann, H. P. (2001). Cheese analogues: A review. *International Dairy Journal, 11*(4–7), 505–515.
3. Barbosa, J., Ferreira, V., & Teixeira, P. (2009). Antibiotic susceptibility of enterococci isolated from traditional fermented meat products. *Food Microbiology, 26*(5), 527–532.
4. Beresford, T. P., Fitzsimons, N. A., Brennan, N. L., & Cogan, T. M. (2001). Recent advances in cheese microbiology. *International Dairy Journal, 11*(4–7), 259–274.
5. Blatchly, R. A., Delen, Z., & O'Hara, P. B. (2014). Making sense of olive oil: Simple experiments to connect sensory observations with the underlying chemistry. *Journal of Chemical Education, 91*(10), 1623–1630.
6. Chaijan, M., Benjakul, S., Visessanguan, W., & Faustman, C. (2005). Changes of pigments and color in sardine (*Sardinella gibbosa*) and mackerel (*Rastrelliger kanagurta*) muscle during iced storage. *Food Chemistry, 93*(4), 607–617.
7. Chinnaswamy, R., & Bhattacharya, K. R. (1983). Studies on expanded rice: Optimum processing conditions. *Journal of Food Science, 48*(6), 1604–1608.
8. Dennis, H., Hunter, D. L., Chang, D., Kim, S., White, J. L., Cho, J. W., & Paul, D. R. (2001). Effect of melt processing conditions on the extent of exfoliation in organoclay-based nanocomposites. *Polymer, 42*(23), 9513–9522.
9. Devanthi, P. V. P., & Gkatzionis, K. (2019). Soy sauce fermentation: Microorganisms, aroma formation, and process modification. *Food Research International, 120*, 364–374.
10. Everett, D. W., & Auty, M. A. (2008). Cheese structure and current methods of analysis. *International Dairy Journal, 18*(7), 759–773.
11. Falkensammer, F., Arnetzl, G. V., Wildburger, A., & Freudenthaler, J. (2013). Color stability of different composite resin materials. *The Journal of Prosthetic Dentistry, 109*(6), 378–383.
12. FarahAhmed, A., Mohamed, Z., & Ismail, M. M. (2011). Determinants of fresh fish purchasing behavior among Malaysian consumers. *Current Research Journal of Social Sciences, 3*(2), 126–131.

13. Farkye, N. Y. (2004). Cheese technology. *International Journal of Dairy Technology, 57*(2–3), 91–98.
14. Feng, L., Tang, N., Liu, R., Gong, M., Wang, Z., Guo, Y., & Chang, M. (2021). The relationship between flavor formation, lipid metabolism, and microorganisms in fermented fish products. *Food & Function, 12*(13), 5685–5702.
15. Ferguous, H., Mehyar, G., & Ibrahim, S. A. (2023). The antimicrobial and antioxidation activities of olive pomace extract in pasteurized white cheese. *Jordan Journal of Agricultural Sciences, 19*(4), 345–364.
16. Fiddler, W., Doerr, R. C., Gates, R. A., & Fox, J. B. (1984). Comparison of chemiluminescent and AOAC methods for determining nitrite in commercial cured meat products. *Journal of the Association of Official Analytical Chemists, 67*(3), 525–528.
17. Flores, M., & Toldra, F. (2011). Microbial enzymatic activities for improved fermented meats. *Trends in Food Science & Technology, 22*(2–3), 81–90.
18. Gao, R., Zheng, Z., Zhou, J., Tian, H., & Yuan, L. (2019). Effects of mixed starter cultures and exogenous L-Lys on the physiochemical and sensory properties of rapid-fermented fish paste using long snout catfish by-products. *LWT, 108*, 21–30.
19. Gelman, A., Glatman, L., Drabkin, V., & Harpaz, S. (2001). Effects of storage temperature and preservative treatment on shelf life of the pond-raised freshwater fish, silver perch (*Bidyanus bidyanus*). *Journal of Food Protection, 64*(10), 1584–1591.
20. Hayaloglu, A. A. (2016). Cheese: Microbiology of cheese. *Reference Module in Food Science, 1*, 1–11.
21. Hernández, H., Nunes, M. C., Prista, C., & Raymundo, A. (2022). Innovative and healthier dairy products through the addition of microalgae: A review. *Foods, 11*(5), 755.
22. Huda, N. (2012). Malaysian fermented fish products. In *Handbook of Animal-Based Fermented Food and Beverage Technology* (pp. 709–715).
23. Jansen, K. M. B., Van Dijk, D. J., & Husselman, M. H. (1998). Effect of processing conditions on shrinkage in injection molding. *Polymer Engineering & Science, 38*(5), 838–846.
24. Jiménez-Colmenero, F. (2007). Healthier lipid formulation approaches in meat-based functional foods: Technological options for replacement of meat fats by non-meat fats. *Trends in Food Science & Technology, 18*(11), 567–578.

25. Kołożyn-Krajewska, D., & Dolatowski, Z. J. (2009). Probiotics in fermented meat products. *Acta Scientiarum Polonorum Technologia Alimentaria, 8*(2), 61–74.
26. Kumar, P., Chatli, M. K., Verma, A. K., Mehta, N., Malav, O. P., Kumar, D., & Sharma, N. (2017). Quality, functionality, and shelf life of fermented meat and meat products: A review. *Critical Reviews in Food Science and Nutrition, 57*(13), 2844–2856.
27. Lawrence, R. C., Heap, H. A., & Gilles, J. (1984). A controlled approach to cheese technology. *Journal of Dairy Science, 67*(8), 1632–1645.
28. Lee, W. J., & Lucey, J. A. (2010). Formation and physical properties of yogurt. *Asian-Australasian Journal of Animal Sciences, 23*(9), 1127–1136.
29. Leroy, F., Verluyten, J., & De Vuyst, L. (2006). Functional meat starter cultures for improved sausage fermentation. *International Journal of Food Microbiology, 106*(3), 270–285.
30. Li, C., Bassey, A. P., & Zhou, G. (2023). Molecular changes of meat proteins during processing and their impact on quality and nutritional values. *Annual Review of Food Science and Technology, 14*(2), 85–111.
31. Liszkowska, W., & Berlowska, J. (2021). Yeast fermentation at low temperatures: Adaptation to changing environmental conditions and formation of volatile compounds. *Molecules, 26*(4), 1035.
32. Little, C. L. (1998). The microbiological quality of ready-to-eat dried and fermented meat and meat products. *International Journal of Environmental Health Research, 8*(4), 277–284.
33. Lu, F., Ding, Y., Ye, X., & Liu, D. (2010). Cinnamon and nisin in alginate–Calcium coating maintains quality of fresh northern snakehead fish fillets. *LWT–Food Science and Technology, 43*(9), 1331–1335.
34. Majerska, J., Michalska, A., & Figiel, A. (2019). A review of new directions in managing fruit and vegetable processing by-products. *Trends in Food Science & Technology, 88*(3), 207–219.
35. Mallia, S., Escher, F., & Schlichtherle-Cerny, H. (2008). Aroma-active compounds of butter: A review. *European Food Research and Technology, 226*(4), 315–325.
36. McMahon, D. J., Fife, R. L., & Oberg, C. J. (1999). Water partitioning in Mozzarella cheese and its relationship to cheese meltability. *Journal of Dairy Science, 82*(7), 1361–1369.
37. McSweeney, P. L. H., Ottogalli, G., & Fox, P. F. (2004). Diversity of cheese varieties: An overview. In *Cheese: Chemistry, Physics and Microbiology* (Vol. 2, pp. 1–23).

38. Meydani, S. N., & Ha, W. K. (2000). Immunologic effects of yogurt. *The American Journal of Clinical Nutrition, 71*(4), 861–872.

39. Milovanovic, B., Djekic, I., Miocinovic, J., Djordjevic, V., Lorenzo, J. M., Barba, F. J., & Tomasevic, I. (2020). What is the color of milk and dairy products and how is it measured? *Foods, 9*(11), 1629.

40. Mistry, V. V. (2001). Low fat cheese technology. *International Dairy Journal, 11*(4–7), 413–422.

41. Morzel, M., Fitzgerald, G. F., & Arendt, E. K. (1997). Fermentation of salmon fillets with a variety of lactic acid bacteria. *Food Research International, 30*(10), 777–785.

42. Mosqueda-Jimenez, D. B., Narbaitz, R. M., Matsuura, T., Chowdhury, G., Pleizier, G., & Santerre, J. P. (2004). Influence of processing conditions on the properties of ultrafiltration membranes. *Journal of Membrane Science, 231*(1–2), 209–224.

43. Munekata, P. E., Pateiro, M., Tomasevic, I., Domínguez, R., da Silva Barretto, A. C., Santos, E. M., & Lorenzo, J. M. (2022). Functional fermented meat products with probiotics—A review. *Journal of Applied Microbiology, 133*(1), 91–103.

44. Neiva, C. R. P., Machado, T. M., Tomita, R. Y., Furlan, É. F., Lemos Neto, M. J., & Bastos, D. H. M. (2011). Fish crackers development from minced fish and starch: An innovative approach to a traditional product. *Food Science and Technology, 31*, 973–979.

45. Nout, M. R. (2001). Fermented foods and their production. In *Fermentation and Food Safety* (pp. 1–19).

46. Olgunoglu, I. A. (2013). Determination of color characterization in some economical freshwater fish species fillets during iced storage for 24 days. *Indian Journal of Animal Research, 47*(3), 187–195.

47. Panda, S. H., Ray, R. C., El Sheikha, A. F., Montet, D., & Worawattanamateekul, W. (2011). Fermented fish and fish products: An overview. *Aquaculture Microbiology and Biotechnology, 2*(1), 132–172.

48. Papavergou, E. J., Savvaidis, I. N., & Ambrosiadis, I. A. (2012). Levels of biogenic amines in retail market fermented meat products. *Food Chemistry, 135*(4), 2750–2755.

49. Rankin, S. A., & Brewer, J. L. (1998). Color of nonfat fluid milk as affected by fermentation. *Journal of Food Science, 63*(1), 178–180.

50. Rhee, S. J., Lee, J. E., & Lee, C. H. (2011). Importance of lactic acid bacteria in Asian fermented foods. In *Microbial Cell Factories, 10*, 1–13.

51. Riebroy, S., Benjakul, S., & Visessanguan, W. (2008). Properties and

acceptability of Som-fug, a Thai fermented fish mince, inoculated with lactic acid bacteria starters. *LWT–Food Science and Technology, 41*(4), 569–580.

52. Ríos, J. M., Tesitore, G., & de Mello, F. T. (2022). Does color play a predominant role in the intake of microplastic fragments by freshwater fish: An experimental approach with *Psalidodon eigenmanniorum*. *Environmental Science and Pollution Research, 29*(32), 49457–49464.
53. Rodríguez, H., Curiel, J. A., Landete, J. M., de las Rivas, B., de Felipe, F. L., Gómez-Cordovés, C., & Muñoz, R. (2009). Food phenolics and lactic acid bacteria. *International Journal of Food Microbiology, 132*(2–3), 79–90.
54. Sangjindavong, M., Chuapoehuk, P., Runglerdkriangkrai, J., Klaypradit, W., & Vareevanich, D. (2008). Fermented fish product (pla-ra) from marine fish and preservation. *Kasetsart Journal (Natural Science), 42*(1), 129–136.
55. Sharma, H., Ozogul, F., Bartkiene, E., & Rocha, J. M. (2023). *Critical Reviews in Food Science and Nutrition, 63*(21), 4819–4841.
56. Sharma, R., Garg, P., Kumar, P., Bhatia, S. K., & Kulshrestha, S. (2020). Microbial fermentation and its role in quality improvement of fermented foods. *Fermentation, 6*(4), 106.
57. Singh, V. P., Pathak, V., & Verma, A. K. (2012). Fermented meat products: Organoleptic qualities and biogenic amines. *American Journal of Food Technology, 7*(5), 278–288.
58. Stadnik, J., & Kęska, P. (2015). Meat and fermented meat products as a source of bioactive peptides. *Acta Scientiarum Polonorum Technologia Alimentaria, 14*(3), 181–190.
59. Suman, S. P., Hunt, M. C., Nair, M. N., & Rentfrow, G. (2014). Improving beef color stability: Practical strategies and underlying mechanisms. *Meat Science, 98*(3), 490–504.
60. Talon, R., Leroy, S., & Lebert, I. (2007). Microbial ecosystems of traditional fermented meat products: The importance of indigenous starters. *Meat Science, 77*(1), 55–62.
61. Tamine, M. (2000). Surface spin-wave spectra in trifluorides MF3 (M= Fe, Cr, V). *Surface Science, 469*(1), 45–54.
62. Tavares, J., Martins, A., Fidalgo, L. G., Lima, V., Amaral, R. A., Pinto, C. A., & Saraiva, J. A. (2021). Fresh fish degradation and advances in preservation using physical emerging technologies. *Foods, 10*(4), 780.
63. Todorov, S. D., Stojanovski, S., Iliev, I., Moncheva, P., Nero, L. A., &

Ivanova, I. V. (2017). Technology and safety assessment for lactic acid bacteria isolated from traditional Bulgarian fermented meat product "lukanka". *Brazilian Journal of Microbiology, 48*, 576–586.

64. Wang, J., & Fung, D. Y. (1996). Alkaline-fermented foods: A review with emphasis on pidan fermentation. *Critical Reviews in Microbiology, 22*(2), 101–138.
65. Wang, Z., Wang, Z., Ji, L., Zhang, J., Zhao, Z., Zhang, R., & Chen, L. (2021). A review: Microbial diversity and function of fermented meat products in China. *Frontiers in Microbiology, 12*, 645435.
66. Williams, J. S. E. (2005). Influence of variety and processing conditions on acrylamide levels in fried potato crisps. *Food Chemistry, 90*(4), 875–881.
67. Wright, A. J., Scanlon, M. G., Hartel, R. W., & Marangoni, A. G. (2001). Rheological properties of milkfat and butter. *Journal of Food Science, 66*(8), 1056–1071.
68. Yu, Z., Su, Y., Zhang, Y., Zhu, P., Mei, Z., Zhou, X., & Yu, H. (2021). Potential use of ultrasound to promote fermentation, maturation, and properties of fermented foods: A review. *Food Chemistry, 357*, 129805.
69. Zhai, X., Shi, J., Zou, X., Wang, S., Jiang, C., Zhang, J., & Holmes, M. (2017). Novel colorimetric films based on starch/polyvinyl alcohol incorporated with roselle anthocyanins for fish freshness monitoring. *Food Hydrocolloids, 69*, 308–317.

CHAPTER 6

GAS FERMENTATION FOR COMMERCIAL BIOFUELS PRODUCTION

LEARNING OBJECTIVES

At the end of this chapter, readers will be able to:

1. Understand the biofuels and gas fermentation concepts.
2. Understand the metabolic parameters and organisms involved in gas fermentation for biofuel production.
3. Identify key factors influencing productivity in gas fermentation processes, such as gas composition and medium formulation, pH, and temperature.
4. Understand the process and principles of gasification.
5. Understand the acetogens' ability to convert and CO into valuable biochemical.
6. Understand the role of hydrogenase enzymes in microbial metabolism and their function in hydrogen production under different conditions.

INTRODUCTORY EXAMPLE

LanzaTech is a prominent company specializing in gas fermentation technology, which converts industrial waste gases into biofuels and chemicals. Utilizing their proprietary microbial fermentation processes, LanzaTech effectively transforms waste gases from steel mills into ethanol, demonstrating a sustainable and efficient method of biofuel production. This approach not only reduces greenhouse gas emissions but also provides an innovative use of industrial byproducts.

The process begins with the gasification of biomass to produce syngas, a mixture of carbon monoxide (CO), carbon dioxide (CO_2), and hydrogen (H_2). This syngas is then introduced into a bioreactor containing specialized microbes, such as *Clostridium ljungdahlii*, which metabolize the gases to produce ethanol. The final step involves the separation and purification of ethanol from the fermentation broth, resulting in a viable biofuel. LanzaTech's successful implementation of this technology at a commercial scale underscores the potential of gas fermentation in creating sustainable energy solutions.

UNIT INTRODUCTION

In the past, biofuels have been produced using sugar substrates from food crops like corn (maize), sugar cane, and sugar beet. Over the past decade, there has been considerable fluctuation in global raw sugar prices. The prices have varied significantly over the years, reaching a peak of US $795/ton in February 2011,that was the highest in three decades. The fluctuation in prices can be linked to the worldwide shortage of sugar and crop failures. Uncertainty has resulted from worries and inquiries over the sustainability of using it as a feedstock for industrial biofuel production (Liew et al., 2016).

This analysis seeks to provide insight into the utilization of syngas and industrial waste gas as raw materials, as well as the growing field of gas fermentation for the production of biofuels and other valuable products. In this segment , the benefits of gas fermentation in contrast to traditional sugar-based fermentation and thermochemical conversions will be explored. Gas fermentation provides a more versatile approach to utilizing various feedstocks for syngas production (Sun et al., 2019).

This study will thoroughly examine the biochemistry, genetics, and energy characteristics of the microorganisms engaged in the bioconversion process. Additionally, it will explore the latest developments in systems synthetic biology and biology, which present promising prospects for enhancing biocatalysts in terms of their ability to produce a wider range of products and improve their overall performance. An analysis will be conducted on important processes including gasification, medium formulation, bioreactor designs, and product recovery. Lastly, the current state of gas fermentation's commercialization will be highlighted, and an outlook will be provided (Nanda et al., 2018).

6.1. OVERVIEW OF BIOFUELS AND GAS FERMENTATION

Learning Objectives:

- Understand the biofuels and gas fermentation concept.

Gas fermentation is an industrial process that utilizes a gaseous feedstock consisting of a combination of methane (), carbon monoxide (CO), carbon dioxide (), and hydrogen (), to enzymatically convert them into a desired output, such as fuels or chemicals.

A gas fermentation method relies on microorganisms capable of metabolizing gaseous carbon sources. Acetogenic bacteria are most recognized for their ability to utilize the acetyl-CoA or Wood-Ljungdahl pathway to capture and transform carbon monoxide (CO), carbon dioxide (CO_2), and hydrogen (H_2) into biomass and other compounds. They can produce valuable substances like ethanol, butanol, and 2,3-butanediol in an environment without oxygen, namely in a fermentation setup (Liew et al., 2013) (Figure 6.1).

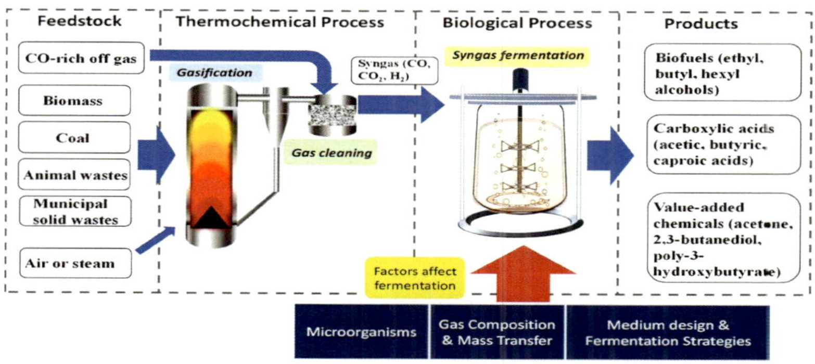

Figure 6.1. *Illustration of syngas fermentation process development for the production of biofuels.*

Source: Ralph S. Tanner, Creative Commons License.

Clostridium ljungdahlii and *C. autoethanogenum* strains are primarily utilized for commercial applications. Additional acetogenic bacteria are currently being developed as organisms for production purposes. There is a huge amount of activity in the fields of synthetic biology and genetic/metabolism engineering to alter these species. In addition, there are efforts to include metabolic pathways into established non-acetogenic organisms such as Escherichia coli or yeasts, in order to broaden the range of possibilities for fermentation processes (Nanda et al., 2023).

In addition, aerobic bacteria can also be utilized for gas fermentation. Aerobic hydrogen oxidizing bacteria, also known as Knall gas bacteria, can convert CO_2 into organic compounds by utilizing H_2 as the source of electrons and O_2 as the final acceptor of electrons. This process is carried out by the Calvin-Benson-Bassham (CBB) cycle. Remarkably, this metabolic process enables the generation of a large amount of organic matter and the creation of more intricate substances, such as poly-hydroxyalkanoates (PHA) biodegradable polymers. The Knall gas model organism *Cupriavidus necator*, previously referred to as *Alcaligenes eutrophus*, is capable of producing single-cell-protein (SCP) and naturally accumulating poly-hydroxybutyrate (PHB) (De Castro, 2007).

Did You Know?

Gas fermentation offers environmental benefits by reducing greenhouse gas emissions through the utilization of waste gases. It contributes to sustainability by converting industrial byproducts into valuable biofuels, thus promoting a circular economy approach.

The pressure on oil supply will intensify as global crude oil stocks deplete and demand rises, especially from developing nations. Crude oil prices fell during the 2007–2010 financial crisis, peaking at US $145 a barrel in July 2008. However, elements including the Middle East's political unrest and the world economies' recovery have caused the price of crude oil to return to $100. Based on the current pace of usage, it is projected that the world's petroleum reserves will be depleted within 50 years (Callegari et al., 2020).

Given the harmful environmental consequences of increasing atmospheric CO_2 levels caused by burning fossil fuels, there is an urgent need of cost-effective and environmentally friendly fuels. In order to solve this issue, many countries have passed laws and put policiesthat support R&D and the commercialization of technology that allows for the production of inexpensive, low-emission fuels with little dependency on fossil carbon (Henstra et al., 2007).

For example, by 2020, member countries of the European Union (EU) must reach a target of using renewable fuel for 10% of their transportation needs. With an average annual growth rate of 15–50%, the use of renewable energies—such as wind, solar, and biofuels—rose significantly between 2005 and 2010. In 2009, renewable energy accounted for around 16% of global final energy consumption (Latif et al., 2014). Various biofuels can be produced from biomass, including solid, liquid, and gaseous forms. These biofuels have the potential to play a significant role in the energy sector. As a percentage of all fuels used worldwide, liquid biofuels accounted for 2.7% of gasoline used in motor vehicles in 2010 (Munasinghe & Khanal, 2010).

The United States (US) is the leading global producer of biofuels, with Brazil and the EU following closely behind. In 2009, the United States and Brazil were responsible for around 85% of the global production of bioethanol, with Europe leading the way in generating about 85% of the world's biodiesel. The value of the liquid biofuels industry has increased dramatically in recent years, leading to a notable jump in the global market. In 2011, it reached a staggering US $83 billion, and experts predict that it will continue to grow, projected to reach US $139 billion by 2021 (Ruan et al., 2019) (Figure 6.2).

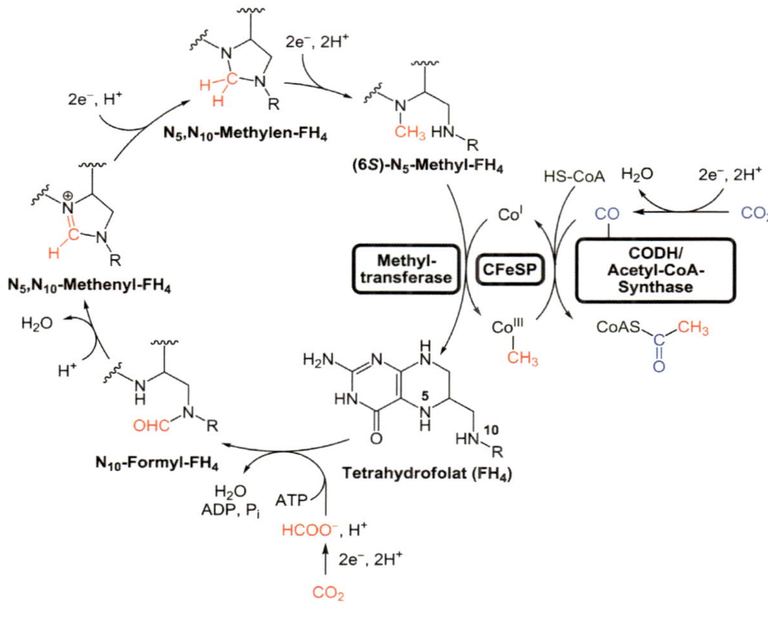

Figure 6.2. *Illustration of the reductive acetyl–CoA pathway.*

Source: Eric Fungmin Liew, Creative Commons License.

Biofuels have a rich history, dating back to the pioneering work of Rudolph Diesel and Nikolaus August Otto. These visionaries recognized the potential of biofuels like ethanol and natural oils as they developed the first Otto cycle combustion and diesel engines. Although the production of ethanol through fermentation has a long history, particularly in brewing beer since ancient times, the discovery of fermentative production of another promising biofuel, butanol, is a relatively recent development that has had a significant impact (Mohammadi et al., 2011).

6.2. GAS FERMENTATION PARAMETERS

Learning Objectives:

- Understand the factors/parameters affecting gas fermentation process.

6.2.1. Gas Composition

The gas fermentation process productivity can be affected by the composition and contaminants of the gas. Due to that fact, H_2 oxidation produces reducing equivalents (rather than CO), a higher molar ratio of H_2: CO enables a more efficient transfer of carbon from CO into products like ethanol. However, hydrogenase is also known to be inhibited by CO, which may have an impact on how well H_2 is used during fermentation (Snowdon, 2001).

H_2 use in *B. methylotrophicum* was suppressed until the depletion of CO. Acetogens can thrive by utilizing carbon dioxide (CO_2) and hydrogen (H_2) when carbon monoxide (CO) is ingested. Typical impurities found in biomass gasification or other waste gases include ash, tar, ethane, char, ethylene, acetylene, hydrogen sulfide (H_2S), ammonia (NH_3), and nitrogen oxide (NO). These factors have been demonstrated to induce cellular dormancy, hinder hydrogen absorption, suppress cell proliferation, and trigger a transition between solventogenesis and acidogenesis in acetogens (Shinohara, 2005).

In the culture media, for example, NH_3 in the feed gas readily changes into NH +. It has been recently discovered that these ions can hinder the activity of hydrogenase and the growth of acetogen "*C. ragsdalei*." There have been various suggestions on how to minimize the impact of impurities. These include implementing 0.025 mm filters, using cyclones or gas scrubbers, and improving gasification efficiency and contaminant removal in the gas flow through the use of substances such as sodium hydroxide, potassium permanganate, or sodium hypochlorite. Acetogens, like *C. ljungdahlii*, are not affected by the presence of H_2S, even at concentrations as high as 5.2% (v/v) (Lu & Lee, 2007).

6.2.2. Substrate Pressure

The microbial growth and product profiles are affected by the partial pressure of syngas components due to the sensitivity of the enzymes associated to substrate exposure. The growth of dense bacterial cell cultures may be hindered by the limited solubility of CO and H_2 in water, which can pose challenges to mass transfer. This problem may be addressed by raising the partial pressure of gaseous substrates (Theodorou et al., 1994).

For example, when the partial pressure of CO (PCO) was increased from 0.35 to 2.0 atm in studies, there was a significant 440% increase in the greatest cell density, a noticeable improvement in ethanol productivity, and a decrease in acetate production in the *C. carboxidivorans* strain P7. In a separate investigation using *C. ljungdahlii*, raising the partial pressure of carbon monoxide (PCO) from 0.8 to 1.8 atm resulted in a beneficial impact on ethanol synthesis. Additionally, the microorganism did not display any inhibition in substrate utilization at elevated PCO levels (Cone et al., 1996).

The Microorganisms that are not CO-tolerant, the impact of increasing the partial pressure of PCO varies. In the case of *Rhodospirillum rubrum*, the effect is minimal. However, for *Peptostreptococcus products* (now known as Blautia products) and *Eubacterium limosum*, it has a negative influence on their doubling time. In a study conducted, it was found that increasing the partial pressure of hydrogen gas (pH 2) to 1,700 mbar had a positive impact on the production of acetate by *A. woodii*. The acetate production rate reached 7.4 g per liter per day (Li et al., 2019).

6.2.3. Medium Formulation

Acetogens can use CO and CO_2/H_2 as the source of carbon and energy. However, they also require other components such as minerals, trace metal elements, vitamins, and reducing agents to maintain a high level of metabolic activity.

Research has shown that the production of ethanol in solventogenic Clostridia is not dependent on growth and that the growth of these bacteria is limited by a decrease in the availability of nitrogen, carbon, and phosphate nutrients. This limitation leads to a change in the equilibrium from acidogenesis to solventogenesis. The ultimate ethanol yield of *C. ljungdahlii* was dramatically boosted to 48 g/l in a CSTR with cell recycling by optimizing the medium formulation. This was achieved by reducing the amounts of B vitamins and eliminating yeast extract. Without cell recycling, the ethanol yield was 23 g/l (Chang et al., 1999).

A separate investigation conducted by Klasson et al. showed that substituting yeast extract with cellobiose resulted in a significant rise in the maximal cell concentration and ethanol output, which was four times higher. An investigation was conducted on the media composition for *C. autoethanogenum* using Plackett-Burman and central composite designs. However, the total ethanol yield waspoor by this process (de Medeiros et al., 2019). Through experimentation, researchers have discovered a more cost-effective and efficient fermentation medium for producing ethanol from syngas. By using 0.5 g/l of cottonseed extract without additional nutrient supplementation, they found that this medium was highly effective for *C. carboxidivorans* strain P7 (Lee et al., 2022).

A recent study demonstrated that increasing the concentrations of specific trace metal ions like , SeO^-, , WO^-, improved the growth, ethanol production, and enzymatic activity (FDH, CODH, and hydrogenase) in "*C. ragsdalei*" under autotrophic conditions. Additionally, the elimination of from the medium also had a positive impact.

Reducing substances such as ascorbic acid, methyl viologen, sodium thioglycolate, titanium (III)-citrate, benzyl viologen, potassium ferricyanide, cysteine-HCl, and sodium sulfide are frequently added to fermentation medium because strict anaerobes require a low redox potential in order to survive. In addition, by enhancing the supply of reducing equivalents for NADH production in the synthesis of alcohol, the introduction of a reducing agent effectively channels the flow of carbon and electrons toward solventogenesis (Losen et al., 2004).

Determining the ideal concentration of reducing agents is crucial because adding too much will slow down microbial growth by lowering the amount of ATP produced via acetogenesis. The sulfur-containing gases found in syngas, such as H_2S, are harmful to chemical catalysts but advantageous to microbiological catalysts because they lower medium redox potential, activate redox-sensitive enzymes like CODH, and encourage the synthesis of alcohol (Azbar et al., 2009).

6.2.4. Medium pH

Similar to other organisms, acetogens require a precise pH range to to reach their full growth potential. Therefore, it is crucial to meticulously regulate the pH of the fermentation medium. The extracellular pH has a direct impact on the intracellular pH, proton motive force, membrane potential, and as a result, the profile of the product and substrate usage.

Reducing the pH medium diverts electron flow and carbon away from acid creation and toward the generation of alcohol in the majority of investigations. With the use of *C. ljungdahlii*, Clausen and Gaddy conducted a two-stage CSTR syngas fermentation system. The pH of the initial reactor was set at 5 to facilitate the growth of cells, whereas the pH of the subsequent reactor was adjusted to a range of 4 to 4.5 to enhance the synthesis of ethanol. Contradictory findings were found in a recent study using *C. ljungdahlii*, wherein ethanol production and cell density were both higher at pH 6.8 than at pH 5.5 (Zhang et al., 2022).

6.2.5. Temperature

In the production of bioethanol, it is important to consider the temperature requirements of acetogens. Mesophilic acetogens perform best at temperatures ranging from 30 to 40°C, while thermophilic acetogens prefer a higher temperature range of 55 to 58°C. The temperature during fermentation has a remarkable impact on various factors, including the growth rate of acetogens, the composition of membrane lipids, the utilization of substrate, and the availability of gas substrate. It is worth noting that gas solubility tends to increase as the temperature decreases. It has been reported that "*C. ragsdalei*" produces more ethanol at 32°C than at 37°C, which is the ideal growth temperature (Phisalaphong et al., 2006).

Did You Know?

Temperature control is crucial in gas fermentation because microbial activity and biofuel production rates are highly sensitive to temperature variations. Optimal temperatures can significantly enhance fermentation efficiency and biofuel yields.

6.3. FEEDSTOCK AND GASIFICATION

Learning Objectives:

- Understand the diverse feedstocks and their preparation for gasification.
- Compare different gasifier types and their operational characteristics.

Almost any carbonaceous material can be utilized as the feedstock for gasification since microorganisms are versatile enough to ferment syngas with a wide range of compositions. Non-food biomass, such as feathers and glycerol, as well as municipal organic wastes, forest residues, agricultural wastes, and some energy crops, can be utilized as gasification feedstock. Biomass is a renewable resource that can be obtained through human or natural processes (e.g., organic wastes) (Mahinpey & Gomez, 2016).

According to estimates, the utilization of modern biomass for energy generation in 2005 was quite low, accounting for only 3.6% of the global energy potential of 250 EJ per year. One notable distinction of utilizing waste streams, like municipal organic waste, for bioethanol production is its economic viability and convenience. These waste materials are readily accessible at affordable prices and are often already collected, reducing the need for additional land use.

On the other hand, non-biomass resources are frequently utilized as feedstocks for the FTP and can also be used for syngas fermentation. Examples of these resources include coal, tar sands, oil shale, cokes, sewage sludge, heavy oil refining wastes, and compressed natural gas.

Certain industries, including oil refining, steel manufacturing, and chemical production, produce significant amounts of waste gas streams that are rich in CO_2 and CO. Utilizing these sources through the process of microbial fermentation effectively transforms harmful waste gas streams into valuable commodities like biofuels. Figure 6.3 shows the gas fermentation process (Qian et al., 2013).

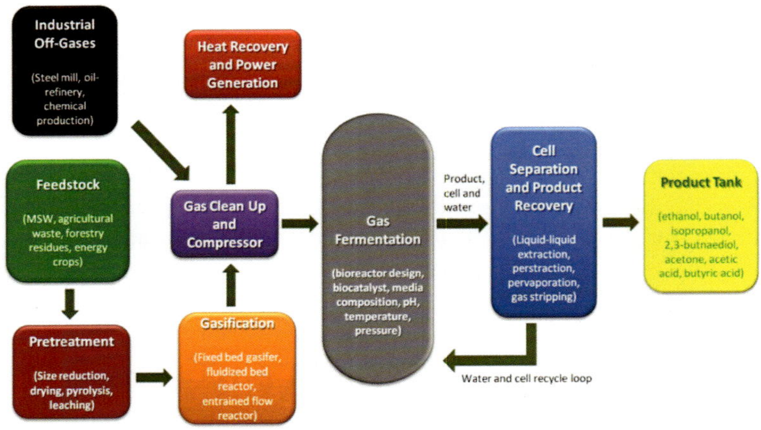

Figure 6.3. *Illustration of an overview of the gas fermentation process.*

Source: Michael Köpke, Creative Commons License.

Before gasification, biomass typically undergoes a pre-treatment process that includes size reduction (such as grinding, chipping, and chopping), drying, fractionation, pyrolysis, and leaching, depending on the gasifier configuration. The pre-treatment process being discussed can lead to significant financial investment and add to the total expense of the biomass feedstock, which can vary from US $16–70 per dry ton (Piatkowski et al., 2011).

Gasification is a thermo-chemical process that converts carbonaceous materials into gaseous intermediates at high temperatures (600–1,000°C) using an oxidizing agent such as steam, air, or oxygen. A mixture of gases including carbon dioxide (CO_2), hydrogen (H_2), carbon monoxide (CO), and nitrogen (N_2) is present in the syngas that is produced. Additionally, there are varying amounts of methane (CH_4), water vapor, and small traces of impurities including carbonyl sulfide (COS), ammonia (NH_3), hydrogen chloride (HCl), hydrogen sulfide (H_2S), nitrogen oxides (NOx), hydrogen cyanide (HCN), light hydrocarbons, phenol, and tar. The presence and quantity of impurities in syngas are influenced by various factors such as the characteristics of the feedstock (such as dust, moisture content, and particle size), the type of gasifier used, and the specific operating conditions (including pressure, temperature, and oxidant) (Lestander et al., 2022).

6.3.1. Fixed Bed Gasifier

Fixed bed gasifiers are further classified as updraft or downdraft reactors based on the direction in which the carbonaceous fuel and oxidant (steam or air) flow. In the counter-current form of the fixed bed gasifier, biomass is fed into the gasifier from the top, while the gasifying chemical is supplied from the bottom. The biomass progresses through different zones within the reactor, each with its specific temperature range. These zones include drying, pyrolysis, gasification, and oxidation. Although this gasifier mode is typically associated with large amounts of tar content in the exit gas, new developments in tar cracking have confirmed very low levels of tar can be achieved (Figure 6.4).

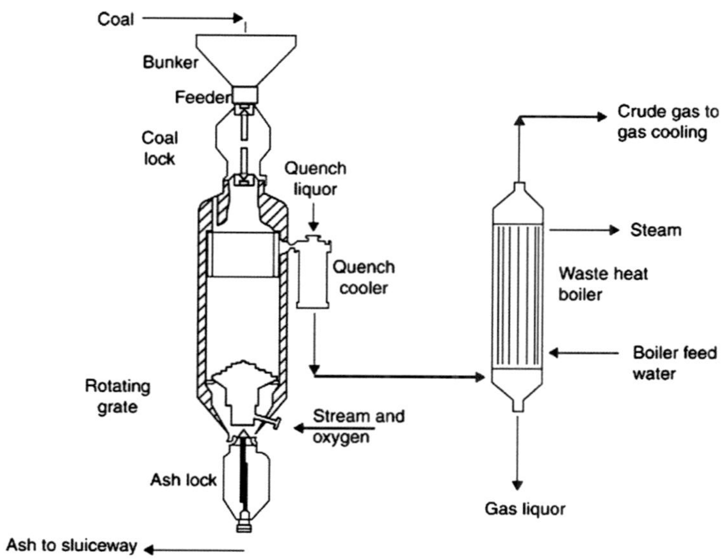

Figure 6.4. *Illustration of fixed bed gasifier.*

Source: Atefeh Akbarian, Creative Commons License.

The efficient exchange of heat between the oxidizing agent and the fuel feed leads to a decrease in gas exit temperature, resulting in improved thermal efficiency. The carbonaceous fuel and oxidizing agent can flow in the same direction in the downdraft (co-current) gasifier because of its construction, which is comparable to that of the updraft reactor. When comparing the updraft gasifier to the downdraft reactor, it is worth noting that the former had reduced amount of tar in the exit gas. However, it does have a lower thermal efficiency. Considering the constraints in the constriction area of the reactor, this type of gasifier has been classified as unsuitable for large-scale operation (Begum et al., 2013).

6.3.2. Fluidized Bed Reactor

In a fluidized bed reactor, a fluidization medium such as air or steam passes through the reactor to combine the carbonaceous fuel with an inert bed material, like silica sand. The improved heat exchange between the fuel components made possible by the use of inert bed results in almost constant operating temperatures and ideal feedstock conversion efficiencies. The bed material's melting point is typically between 800°C and 900°C, commonly sets a maximum operating temperature limit for the gasifier.

Additionally, fluidized bed reactors are ideal options for scaling up due to the reactor's unique mixing qualities and geometry. Given its desirable properties, the fluidized bed reactor has become the preferred choice for gasifying biomass feedstock. Unfortunately, this particular gasifier model is not well-suited for feedstocks that contain high levels of alkali metals and ash. The existence of these constituents results in adhesiveness and the creation of bigger aggregates, ultimately causing disruption to the seamless flow of the reactor (Singh et al., 2022) (Table 6.1).

Table 6.1. Composition of Syngas and Other Gas Streams from Different Sources

	Composition vol%, Dry Basis					
	CO_2	**CO**				**Other**
Non-Biomass Source						
Coal gasification	10	59.4	0.6	29.4	0	0.6
Partial oxidation of heavy fuel oil	4.3	47	1.4	46	0.3	1
Coke oven gas	1.4	5.6	4.3	55.4	28.4	4.9
Steam reforming of natural gas	8.1	15.5	0.2	75.7	0.5	0

Hardwood chips + 20 wt.% liquid crude glycerol	11.67	19.73	NR*	19.38	3.82	NR*
Water gas	3.4	30	13.1	31.7	12.2	9.6
Steam reforming of naphtha	15.8	6.7	2.6	65.9	6.3	2.7
Steel mill	22	44	32	2	0	0
Biomass and Organic Waste Sources						
Dairy biomass	15.7	8.7	56	18.6	0.6	0.4
Cacao shell	16.02	8	61.45	9.02	2.34	3.17
Demolition wood + sewage sludge	15.02	10.53	60.46	8.02	3.19	2.78
Kentucky bluegrass straw	17.4	12.9	64.2	2.6	2.1	0.8
Switchgrass	16.5	14.7	56.8	4.4	4.2	3.4
Willow	17.1	9.4	60.42	7.2	3.3	2.58

Source: Séan D. Simpson, Creative Commons License.

6.3.3. Entrained Flow Reactor

The entrained flow reactor is a popular option for gasifying coal, pet coke, and refinery wastes on a large scale because of its low tar generation and high carbon conversion efficiency. There is no any requirement for inert bed material for this gasifier to function. Rather, it depends on combining the oxidizing agent and feedstocks at a high speed to provide a pneumatic transport regime.

This process can change tars and methane when the temperature exceeds 1,200–1,500°C, improving the quality of syngas. It is crucial to note that this technology necessitates the need for feedstocks to be finely pulverized into particles of approximately 50 μm before they can be used. While this is not a significant problem for coal, it poses considerable challenges and expenses when it comes to biomass sources (Kolb et al., 2016).

6.4. BIOCHEMISTRY AND MICROBES OF GAS FERMENTATION

Learning Objectives:

- Understand biochemical pathways and enzymes involved in CO_2 fixation by acetogens.

As obligate anaerobes, acetogens transform CO_2 into the acetyl part of acetyl-coenzyme A (CoA) via the reductive acetyl-CoA pathway. This process serves to conserve energy and incorporate CO_2 into cellular carbon. There are several recognized biological mechanisms that lead to full autotrophic fixation of CO_2. These consist of the 3-hydroxypropionate/malyl-CoA cycle, the reductive tricarboxylic acid (TCA) cycle, the Calvin cycle, and the 3-hydroxypropionate/4-hydroxybutyrate cycle.

It has been proposed that the acetyl-CoA route may have been the first autotrophic activity on Earth because of the anoxic early Earth's atmosphere and the pipeline's biochemical simplicity in comparison to other autotrophic pathways. The 'Wood-Ljungdahl' pathway, which is the reductive acetyl-CoA pathway, is named after Harland G. Wood and Lars G. Ljungdahl, who conducted substantial research on the pathway's enzymatic and chemical features. *Moorella thermoacetica*, formerly known as *Clostridium thermoaceticum*, or the CODH/ACS pathway—named for the pathway's major enzyme, carbon monoxide dehydrogenase/acetyl-CoA synthase—was used in their research.

This ancient pathway is found in a wide range of bacterial genera, including *Acetoanaerobium*, *Acetitomaculum*, *Acetonema*, *Acetohalobium*, *Acetobacterium*, *Alkalibaculum*, "*Butyribacterium*," "*Bryantella*," *Clostridium*, *Caloramator*, *Holophaga*, *Moorella*, *Eubacterium*, *Natronincola*, *Natroniella*, *Oxobacter*, *Ruminococcus*, *Sporomusa*, *Syntrophococcus*, *Thermoanaerobacter*, *Thermoacetogenium*, *Tindallia*, and *Treponema*. Acetogens have the remarkable ability to use gases such as $CO_2 + H_2$, and/or CO and convert them into valuable products like acetic acid and ethanol. This process follows specific stoichiometries (Phillips et al., 2017):

$$2\,CO_2 + 4\,H_2 \Rightarrow CH_3COOH + 2\,H_2O \qquad \Delta H = -75.3\ kJ/mol \tag{6.1}$$

$$2\,CO_2 + 6\,H_2 \Rightarrow C_2H_5OH + 3\,H_2O \qquad \Delta H = -97.3\ kJ/mol \tag{6.2}$$

$$4\,CO + 2\,H_2O \Rightarrow CH_3COOH + 2\,CO_2 \qquad \Delta H = -154.9\ kJ/mol \tag{6.3}$$

$$6CO + 3\,H_2O \Rightarrow C_2H_5OH + 4\,CO_2 \qquad \Delta H = -217.9\ kJ/mol \tag{6.4}$$

A key mechanism that effectively adds CO_2 to biomass by terminal electron-acceptance is the acetyl-CoA pathway. There are mainly two branches, one Eastern (or Carbonyl) and one Western (or Methyl) (Figure 6.5). A group of enzymes in the Western branch decreases CO_2 by six electrons to the methyl group of acetyl-CoA. At the beginning of this process,dehydrogenase transforms CO_2 into formate.

Formyl-H_4 folate synthase catalyzes the condensation of formate with H_4 folate, resulting in the formation of 10-formyl-H_4 folate. This compound is subsequently transformed to 5,10-methenyl-H_4 folate by a cyclohydrolase enzyme. Next, a dehydrogenase enzyme converts methenyl- to 5,10-methylene-H_4 dihydrofolate. This is subsequently converted by methylene-H_4 folate reductase into (6S)-5-CH_3–H_4 folate.

An essential enzyme called B12-dependent methyltransferase (MeTr) transfers a methyl group from (6S)-5--folate to a protein inside the bi-functional carbon monoxide dehydrogenase/acetyl-CoA synthase (CODH/ACS) complex known as corrinoid iron-sulfur protein (CoFeSP). The bi-functional CODH/ACS enzyme complex is composed of two independent proteins: a $\gamma\delta$ heterodimer (CoFeSP) and an $\alpha_2\beta_2$ tetramer (CODH/ACS). These proteins are frequently organized in an operon, along with MeTr (Griffin & Schultz, 2012).

In the Eastern branch, the CODH component catalyzes the reduction of CO_2 to CO. The central molecule, acetyl-CoA, is finally generated when CO, methyl group (bound to CoFeSP) and CoASH are condensed by ACS. Given the pivotal role of CODH/ACS, it is unsurprising that this complex was found to be the most highly expressed transcripts under autotrophic conditions in C. auto ethanogenum, and can represent up to 2% of the soluble cell protein of an acetogen. CODH/ACS is not unique to acetogenic bacteria, as it is also present in sulfate-reducing bacteria, desulfitobacteria, and Archaea (methanogens and Archaeoglobus.

The reducing equivalents required for the fixation of carbon into acetyl-CoA come from the oxidation of molecular hydrogen under chemolithoautotrophic growth, or NADH and reduced ferredoxin under heterotrophic growth. (Wood, 1981).

Practice Problem

If a particular microbial strain converts 85% of the carbon monoxide (CO) in the feed gas to ethanol, and the feed gas contains 500 moles of CO, calculate the moles of ethanol produced.

Solutions to Practice Problem

Moles of CO converted = 85% of 500 moles
Moles of CO converted = $0.85 \times 500 = 425$ moles
Assuming 1 mole of CO produces 1 mole of ethanol:
Moles of ethanol produced = 425 moles.

The hydrogenase process proceeds in a certain direction depending on the redox potential of the materials that can interact with the enzyme. Hydrogen evolution is taking place when there is an available electron donor, while the presence of an electron acceptor leads to hydrogen oxidation. There are three different classes of metalloenzymes called hydrogenases: $[FeFe]^-$, $[NiFe]^-$, and $[Fe]^-$hydrogenases. These classes are classified based on their phylogenetic distinctions. The Ech hydrogenase in *Methanosarcina barkeri* has been shown to convert H_2 into reduced ferredoxin. In the process of acetoclastic methanogenesis, Ech hydrogenase facilitates the oxidation of ferredoxin, resulting in the production of H_2.

Fun fact:

One of the microbes commonly used in gas fermentation, Clostridium ljungdahlii, can produce ethanol directly from syngas. This bacterium has evolved a unique metabolic pathway that enables it to convert gases like CO and H into ethanol and acetate.

Despite the discovery of Ech-like hydrogenase in *C. phytofermentans*, *C. thermocellum*, *C. cellulolyticum*, and *C. papyrosolvens* through genome analysis, the specific functions of these enzymes have yet to be determined. Clostridia possess various types of [FeFe]-hydrogenases, indicating their capacity to adapt quickly to different environmental circumstances. The [FeFe]-hydrogenase of *C. pasteurianum* has been extensively researched and is considered one of the well-studied enzymes in its class. It enables the movement of electrons from ferredoxins or flavodoxins that have gained electrons to protons, leading to the creation of H_2 (Achinas et al., 2020).

It has been hypothesized that a trimeric [FeFe]-hydrogenase, which has been discovered in *C. beijerinckii*, *C. difficile*, and *C. carboxidivorans*, can convert formate oxidation into the production of H_2 by reducing protons. An interesting discovery was made in *Thermotoga maritima*, where researchers identified a unique trimetric [FeFe]-hydrogenase. This enzyme has the remarkable ability to simultaneously oxidize reduced ferredoxin and NADH, resulting in the evolution of hydrogen at low H_2 partial pressure. The authors proposed that under conditions of high H_2 partial pressure, NADH is oxidized to generate ethanol.

Through computational analysis, it was discovered that certain strains of *Clostridia*, such as *C. beijerinckii* and *C. thermocellum*, possess homologs of this bifurcating hydrogenase. Furthermore, CODH/ACS, hydrogenases, and pyruvate: ferredoxin oxidoreductase (PFOR) from *M. thermoacetica* have been found to possess the ability to produce hydrogen. This could serve as a way to eliminate surplus-reducing equivalents in situations where electron carriers are scarce and there is a sufficient concentration of CO to hinder conventional hydrogenases (Daniell et al., 2016).

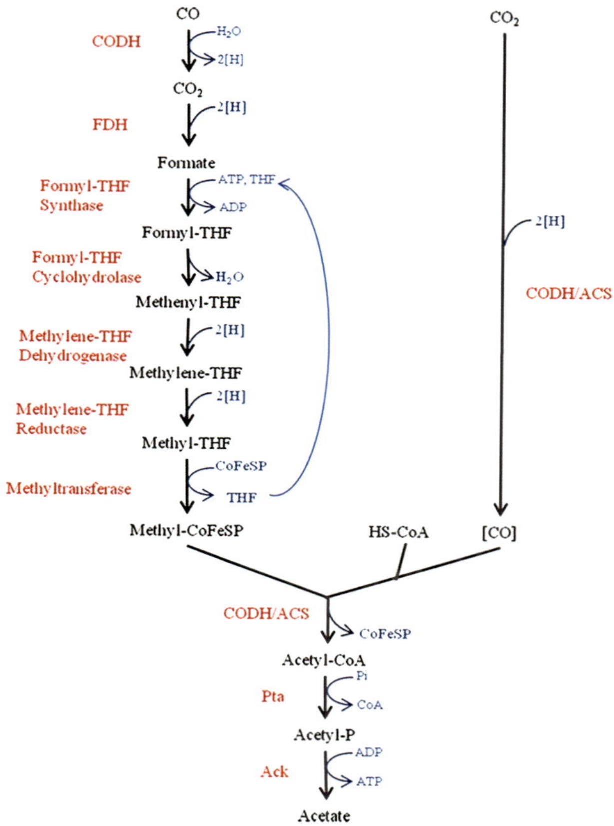

Figure 6.5. *Illustration of Wood-Ljungdahl pathway. [ACS: Acetyl-CoA synthase; Ack: acetate kinase; CODH: carbon monoxide dehydrogenase; CoFeSP: corrinoid iron-sulfur protein; Pta: phosphotransacetylase; THF: tetrahydrofolate; and FDH: formate dehydrogenase].*

Source: Michael Köpke, Creative Commons License.

SUMMARY

- This chapter provides summary of the technological and biochemical processes used to convert gases into biofuels.
- The chapter explores important factors including substrate pressure, gas composition, medium formulation, pH, and temperature, emphasizing their impact on microbial activity and the efficiency of biofuel production.
- The chapter explores different types of gasification reactors, including fluidized bed reactors, fixed bed gasifiers, and entrained flow reactors. It explores their functions in the conversion of feedstocks such as industrial waste gases and biomass-derived syngas into biofuels.
- In addition, it discusses the significance of specific microorganisms and their biochemical processes. The chapter demonstrates how these organisms facilitate the transformation of gases into ethanol and other biofuels.

REVIEW QUESTIONS

1. Primary components and steps in gas fermentation for biofuel production?
2. Impact of gas composition on gas fermentation efficiency? Provide examples.
3. Importance of substrate pressure in gas fermentation and its effects on microbial activity and biofuel yield.
4. Role of medium formulation in gas fermentation and key considerations in designing the fermentation medium.
5. Impact of pH on gas fermentation processes and microbial health.
6. Significance of temperature control in gas fermentation and its effects on the process.
7. Comparison of fixed bed gasifiers, fluidized bed reactors, and entrained flow reactors in gasification.
8. Process of feedstock gasification, types of feedstocks used, and their preparation.
9. Biochemical pathways used by microbes during gas fermentation and their role in biofuel production.

MULTIPLE CHOICE QUESTIONS

1. **What is the primary focus of the chapter "Gas Fermentation for Commercial Biofuels Production"?**
   a. Chemical synthesis of biofuels
   b. Overview of gas fermentation processes and parameters for biofuel production
   c. Agricultural methods for biofuel crops
   d. Economic analysis of biofuel markets
2. **Which microbial process is crucial for gas fermentation in biofuel production?**
   a. Photosynthesis
   b. Glycolysis
   c. Biochemical pathways specific to gas fermentation
   d. Protein synthesis
3. **What advantage does gas fermentation offer in the production of biofuels?**
   a. Higher energy consumption
   b. Utilization of diverse feedstock's
   c. Lower efficiency compared to traditional methods
   d. Limited environmental benefits
4. **In gas fermentation, what role do microbes play?**
   a. They catalyze the conversion of gases into biofuels
   b. They consume the final biofuel product
   c. They provide the substrate for gasification
   d. They are used to maintain reactor temperature
5. **What is one key benefit of gas fermentation for biofuel production highlighted in the chapter?**
   a. High operational costs
   b. Limited microbial strain availability
   c. Requirement for high purity gases
   d. Ability to use a wide range of feedstocks

Answers to Multiple Questions

1. (b); 2. (c); 3. (b); 4. (a); 5. (d)

REFERENCES

1. Achinas, S., Achinas, V., & Euverink, G. J. W. (2020). Microbiology and biochemistry of anaerobic digesters: An overview. In *Bioreactors* (pp. 17–26).
2. Azbar, N., Dokgöz, F. T. Ç., & Peker, Z. (2009). Optimization of basal medium for fermentative hydrogen production from cheese whey wastewater. *International Journal of Green Energy, 6*(4), 371–380.
3. Begum, S., Rasul, M. G., Akbar, D., & Ramzan, N. (2013). Performance analysis of an integrated fixed bed gasifier model for different biomass feedstocks. *Energies, 6*(12), 6508–6524.
4. Callegari, A., Bolognesi, S., Cecconet, D., & Capodaglio, A. G. (2020). Production technologies, current role, and future prospects of biofuels feedstocks: A state-of-the-art review. *Critical Reviews in Environmental Science and Technology, 50*(4), 384–436.
5. Chang, I. S., Kim, B. H., Kim, D. H., Lovitt, R. W., & Sung, H. C. (1999). Formulation of defined media for carbon monoxide fermentation by Eubacterium limosum KIST612 and the growth characteristics of the bacterium. *Journal of Bioscience and Bioengineering, 88*(6), 682–685.
6. Cone, J. W., van Gelder, A. H., Visscher, G. J., & Oudshoorn, L. (1996). Influence of rumen fluid and substrate concentration on fermentation kinetics measured with a fully automated time related gas production apparatus. *Animal Feed Science and Technology, 61*(1–4), 113–128.
7. Daniell, J., Nagaraju, S., Burton, F., Köpke, M., & Simpson, S. D. (2016). Low-carbon fuel and chemical production by anaerobic gas fermentation. *Anaerobes in Biotechnology, 7*(5), 293–321.
8. De Castro, J. F. (2007). Biofuels: An overview. Final report. Prepared for DGIS/DMW/IB. Environmental infrastructure and impact division, environment and water department, directorate-general for international cooperation (DGIS), the Hague, Netherlands. *8*(4), 6548–6550.
9. de Medeiros, E. M., Posada, J. A., Noorman, H., & Filho, R. M. (2019). Dynamic modeling of syngas fermentation in a continuous stirred-tank reactor: Multi-response parameter estimation and process optimization. *Biotechnology and Bioengineering, 116*(10), 2473–2487.
10. Griffin, D. W., & Schultz, M. A. (2012). Fuel and chemical products from biomass syngas: A comparison of gas fermentation to thermochemical conversion routes. *Environmental Progress & Sustainable Energy, 31*(2), 219–224.

11. Henstra, A. M., Sipma, J., Rinzema, A., & Stams, A. J. (2007). Microbiology of synthesis gas fermentation for biofuel production. *Current Opinion in Biotechnology, 18*(3), 200–206.

12. Kolb, T., Aigner, M., Kneer, R., Mueller, M., Weber, R., & Djordjevic, N. (2016). Tackling the challenges in modelling entrained-flow gasification of low-grade feedstock. *Journal of the Energy Institute, 89*(4), 485–503.

13. Latif, H., Zeidan, A. A., Nielsen, A. T., & Zengler, K. (2014). Trash to treasure: Production of biofuels and commodity chemicals via syngas fermenting microorganisms. *Current Opinion in Biotechnology, 27*, 79–87.

14. Lee, M., Jang, N., Kang, B., Yasin, M., Kim, J. Y., & Chang, I. S. (2022). Proposition of cost-effective industrial grade medium for synthesis gas fermentation through evaluation of economic feasibility. *Bioresource Technology Reports, 19*, 101208.

15. Lestander, T. A., Weiland, F., Grimm, A., Rudolfsson, M., & Wiinikka, H. (2022). Gasification of pure and mixed feedstock components: Effect on syngas composition and gasification efficiency. *Journal of Cleaner Production, 369*, 133330.

16. Li, L., Wang, Y., & Li, Y. (2019). Effects of substrate concentration, hydraulic retention time and headspace pressure on acid production of protein by anaerobic fermentation. *Bioresource Technology, 283*, 106–111.

17. Liew, F. M., Köpke, M., & Simpson, S. D. (2013). Gas fermentation for commercial biofuels production. *Liquid, Gaseous and Solid Biofuels-Conversion Techniques, 4*(3), 125–173.

18. Liew, F., Martin, M. E., Tappel, R. C., Heijstra, B. D., Mihalcea, C., & Köpke, M. (2016). Gas fermentation—A flexible platform for commercial scale production of low-carbon-fuels and chemicals from waste and renewable feedstocks. *Frontiers in Microbiology, 7*, 694.

19. Losen, M., Frölich, B., Pohl, M., & Büchs, J. (2004). Effect of oxygen limitation and medium composition on *Escherichia coli* fermentation in shake-flask cultures. *Biotechnology Progress, 20*(4), 1062–1068.

20. Lu, Y., & Lee, T. (2007). Influence of the feed gas composition on the Fischer-Tropsch synthesis in commercial operations. *Journal of Natural Gas Chemistry, 16*(4), 329–341.

21. Mahinpey, N., & Gomez, A. (2016). Review of gasification fundamentals and new findings: Reactors, feedstock, and kinetic studies. *Chemical Engineering Science, 148*, 14–31.

22. Mohammadi, M., Najafpour, G. D., Younesi, H., Lahijani, P., Uzir, M. H., & Mohamed, A. R. (2011). Bioconversion of synthesis gas to second

generation biofuels: A review. *Renewable and Sustainable Energy Reviews, 15*(9), 4255–4273.

23. Munasinghe, P. C., & Khanal, S. K. (2010). Biomass-derived syngas fermentation into biofuels: Opportunities and challenges. *Bioresource Technology, 101*(13), 5013–5022.
24. Nanda, S., Pattnaik, F., Patra, B. R., Kang, K., & Dalai, A. K. (2023). A review of liquid and gaseous biofuels from advanced microbial fermentation processes. *Fermentation, 9*(9), 813.
25. Nanda, S., Rana, R., Sarangi, P. K., Dalai, A. K., & Kozinski, J. A. (2018). A broad introduction to first-, second-, and third-generation biofuels. In *Recent Advancements in Biofuels and Bioenergy Utilization* (Vol. 7, pp. 1–25).
26. Phillips, J. R., Huhnke, R. L., & Atiyeh, H. K. (2017). Syngas fermentation: A microbial conversion process of gaseous substrates to various products. *Fermentation, 3*(2), 28.
27. Phisalaphong, M., Srirattana, N., & Tanthapanichakoon, W. (2006). Mathematical modeling to investigate temperature effect on kinetic parameters of ethanol fermentation. *Biochemical Engineering Journal, 28*(1), 36–43.
28. Piatkowski, N., Wieckert, C., Weimer, A. W., & Steinfeld, A. (2011). Solar-driven gasification of carbonaceous feedstock—A review. *Energy & Environmental Science, 4*(1), 73–82.
29. Qian, K., Kumar, A., Patil, K., Bellmer, D., Wang, D., Yuan, W., & Huhnke, R. L. (2013). Effects of biomass feedstocks and gasification conditions on the physiochemical properties of char. *Energies, 6*(8), 3972–3986.
30. Ruan, R., Zhang, Y., Chen, P., Liu, S., Fan, L., Zhou, N., & Li, B. (2019). Biofuels: Introduction. In *Biofuels: Alternative Feedstocks and Conversion Processes for the Production of Liquid and Gaseous Biofuels* (Vol. 3, pp. 3–43). Academic Press.
31. Shinohara, H. (2005). A new technique to estimate volcanic gas composition: Plume measurements with a portable multi-sensor system. *Journal of Volcanology and Geothermal Research, 143*(4), 319–333.
32. Singh, M., Salaudeen, S. A., Gilroyed, B. H., & Dutta, A. (2022). Simulation of biomass-plastic co-gasification in a fluidized bed reactor using Aspen plus. *Fuel, 319*, 123708.
33. Snowdon, L. R. (2001). Natural gas composition in a geological environment and the implications for the processes of generation and preservation. *Organic Geochemistry, 32*(7), 913–931.

34. Sun, X., Atiyeh, H. K., Huhnke, R. L., & Tanner, R. S. (2019). Syngas fermentation process development for production of biofuels and chemicals: A review. *Bioresource Technology Reports, 7*, 100279.
35. Theodorou, M. K., Williams, B. A., Dhanoa, M. S., McAllan, A. B., & France, J. (1994). A simple gas production method using a pressure transducer to determine the fermentation kinetics of ruminant feeds. *Animal Feed Science and Technology, 48*(3–4), 185–197.
36. Wood, W. A. (1981). Basic biology of microbial fermentation. *Trends in the Biology of Fermentations for Fuels and Chemicals, 7*, 3–17.
37. Zhang, M., Liang, G., Zhang, X., Lu, X., Li, S., Wang, X., & Jiao, P. (2022). The gas production, ruminal fermentation parameters, and microbiota in response to *Clostridium butyricum* supplementation on in vitro varying with media pH levels. *Frontiers in Microbiology, 13*, 960623.

CHAPTER 7

FERMENTED CEREAL AND LEGUME PRODUCTS

LEARNING OBJECTIVES

At the end of this chapter, readers will be able to:

1. Comprehend the diverse types of fermented cereal and legume products.
2. Understand the key microorganisms involved in cereal and legume fermentation.
3. Describe the chemical changes induced by pre-fermentation processes.
4. Explain the chemical changes occurring during fermentation of carbohydrates, proteins, and lipids.
5. Explore the effects of fermentation on mineral and vitamin content in food.
6. Recognize the production of volatile compounds and their impact on flavor in fermented foods.

INTRODUCTORY EXAMPLE

Consider the journey of a small artisanal bakery, "Cultured Crusts," located in the heart of a bustling city. Inspired by traditional baking techniques and a passion for wholesome ingredients, Cultured Crusts specializes in crafting sourdough bread using time-honored fermentation methods. Each loaf is meticulously handcrafted, starting with locally sourced organic flour and pure spring water.

The magic begins with the creation of a sourdough starter, a symbiotic culture of wild yeast and *lactobacillus* bacteria. This living culture, nurtured and fed daily, imparts complex flavors and leavening power to the bread. As the sourdough starter matures, it becomes a vital part of Cultured Crusts' artisanal process, adding depth and character to each batch of dough.

Customers flock to Cultured Crusts not only for the unparalleled taste of their sourdough bread but also for the knowledge that they are consuming a product crafted with care and respect for tradition. As they savor each bite, they become part of a culinary legacy that stretches back through generations, celebrating the artistry and science of fermentation.

UNIT INTRODUCTION

Fermenting different kinds of dietary raw materials has been recognized as a human tradition since ancient times. Cereals, such as maize, rye, rice, wheat, barley, sorghum, oats, and millet, have been subjected to fermentation processes to produce a variety of goods in different regions across the globe.

Fermentation offers numerous benefits to food, such as introducing new flavors, tastes, textures, and aromas. It helps in preserving food, enhancing its nutritional value by improving digestibility and increasing vitamin production, eliminating toxic substancesand reducing cooking time and energy consumption. Therefore, the process of the fermentation brings about changes in the shape, texture, and flavor of food. It also enhances its nutritional content, and ensures its safety (Adebo et al., 2022). The family *Poaceae* includes various cereal grains such as wheat (*Triticum sativum*), pearl millet (*Pennisetum glaucum*), rye (*Secale cereale*), maize (*Zea mays*), barley (*Hordeum vulgare*), rice (*Oryza sativa*), finger millet (*Eleusine coracana*), oats (*Avena sativa*), sorghum (*Sorghum bicolor*), and teff (*Eragrostis tef*).

Cereals play a crucial role in the diets of economically less-developed regions, where the fermentation products derived from cereal grains are widely consumed as staple foods. The production of fermented products, which are derived from various cereal grains, is influenced by the geographic distribution of these grains (Reddy et al., 1983).

Various regions around the world have their preferred fermented products. In Africa and Asia, sorghum and millet-based fermented products are popular, while maize-based ones dominate in Central and South America and Africa. Rice-based fermented products are commonly consumed in China, Southeast Asia, India and the Far East. Wheat-based fermented products hold significance in Turkey, the Middle East, and the Far East. Fermentation products can be referred to by various names depending on the country. The sorghum beer known by different names in various countries is a popular traditional beverage in West Africa. It is consumed in Cameroon, Chad, Nigeria, Burkina Faso, Ghana, and the Ivory Coast (Chavan et al., 1989).

Did You Know?

Sourdough bread has been around for thousands of years, with evidence suggesting that ancient Egyptians were already making fermented bread over 5,000 years ago.

The *Leguminaceae* family, sometimes known as Fabaceae, has various varieties of beans and peas. It ranks second in significance to human nutrition, following the grass family. Cereal grains are significant providers of energy and carbs, whereas legumes are crucial suppliers of proteins. Most of the fermented products have been studied in depth come from soybeans (Glycine max). But fermentation is also used on a variety of legumes (Reddy et al., 1982).

Fermented soya products encompass a wide range of items such as tempeh, cheonggukjang, tofu, soybean sauce, miso, chunjang, doenjang, gochujang, natto, tauchu, tamari, yellow soybean paste and doubanjiang. Additional examples of fermented legume products are dhokla, imriti, papad, idli, dosa, and wadi. Regarding grains, legumes can be fermented either in their natural state or after undergoing the process of malting. Legume fermentation is mostly conducted in South-east Asia and the Far East, including countries such as Korea, Thailand, China, Japan, Taiwan, Indonesia, Philippines, Malaysia, Burma, and Singapore. It is also practiced to a lesser degree in Africa and South America.

Fermentation is commonly practiced in various regions using a range of mixed starter and grain cultures, resulting in variations in the quality of the end product. This can affect its taste, flavor, appearance, safety, and nutritional value. On the other hand, countries that use advanced biotechnology rely on specific fermentation and cultivar cultures to create commercial products with consistent and reliable quality (Achi & Ukwuru, 2015). The chapter delves into the wide variety of fermented foods made from cereal grains and legumes, as well as the complex procedures which used to produce them. It explores various fermented cereal and legume products, providing a comprehensive analysis of their manufacturing techniques. In addition, it discusses the important role of microorganisms used in fermentation and explains the chemical transformations that take place in fermented foods (Adebo et al., 2021).

These changes involve modifications caused by both pre-fermentation processes and fermentation itself, which include conversions in carbohydrates, proteins, and lipids, as well as the creation of volatile chemicals. It provides a thorough comprehension of the complex processes and changes during the fermentation of cereal grains and legumes, which play a role in the creation of a diverse range of culturally important and nutritionally enhanced food products.

7.1. TYPES OF FERMENTED CEREAL PRODUCTS AND THEIR MANUFACTURE

Learning Objectives:

- Understand types and processes of fermented cereal products, including breads and beverages.

There are three main types of cereal fermentation. First comes lactic acid fermentation, which is followed by alcoholic fermentation made by various yeasts including baker's yeast (*Saccharomyces* cerevisiae). Lastly, there is acetic acid fermentation, which is performed by Acetobacter and involves the conversion of alcohol to acetic acid in an atmosphere of excess oxygen. There are various categories of fermented cereal products, including leavened breads, gruels and porridges, and beverages (both alcoholic and nonalcoholic).

By fermenting inside the dough, baker's yeast is used to leaven many kinds of bread. The leavening agent produced by this fermentation is ethanol, which evaporates during baking, and carbon dioxide (Blandino et al., 2003).

There is a wide variety of breads made around the world. For example, there are popular wheat loaves and wheat-based grain mixes that are consumed all over the world. In Central American countries, tortillas made of maize and wheat are popular. The foods of Italy and other southern European nations include pizza, focaccia, ciabatta, and pita bread. Scandinavian countries have crisp breads made of barley, rye, and wheat. In North Africa and Middle East countries, baladi and other flatbreads are commonly consumed. To evaluate these breads, key factors to be considered such as are their volume, texture, and ability to stay fresh for an extended period. It is also crucial to ensure that they meet the necessary standards for microbiological safety (Abou-Zeid, 2016).

Did You Know?

Kimchi, a traditional Korean side dish made from fermented vegetables such as cabbage and radishes, is not only delicious but also rich in probiotics, vitamins, and antioxidants. It is often enjoyed as a spicy and tangy accompaniment to meals.

When it comes to making bread like sourdough, particularly rye bread, a unique combination of yeasts (typically Candida and *Saccharomyces*) work together with *Lactobacillus* to create that distinct sour taste from lactic acid. This process of making sourdough bread involves combining a small portion of previously fermented dough with fresh flour and water. Based on the duration of fermentation, certain bacteria, such as *Lactobacillus*, are present alongside yeast and play a role in enhancing the taste and flavor of bread (Ashaolu & Reale, 2020). Three primary categories of sourdough can be identified:

- Type I of sourdough is characterized by a pH range of 3.8–4.5 and a fermentation temperature range of 20°C–30°C. In this type, the dough is leavened by *Saccharomyces exiguus*, while sourness is produced by *Lactobacillus* species, primarily *L. sanfranciscensis* and *L. pontis*.
- In Type II sourdoughs, the pH is below 3.5, and the fermentation temperature ranges from 30°C to 50°C for a veral days. In this type of fermentation, *Saccharomyces* cerevisiae is responsible for leavening the dough, while *L. pontis* and *L. panis* are the dominant lactic acid-producing bacteria.
- Type III sourdoughs are commonly used in bread-making to enhance the flavor and taste right away (Vogel & Ehrmann, 2008).

Additional bacteria, such as *L. fructivorans*, *L. brevis*, *L. fermentum*, and *L. paralimentarius*, can also be found in these doughs. Occasionally, there may be other bacteria present in sourdough that produce gas, such as hetero fermenting lactic acid bacteria (LAB).

Besides these traditional sourdough breads, some variations incorporate different ingredients. For example, there is the Amish friendship bread, which ferments quickly and includes milk and sugar in its sourdough starter. Additionally, baking soda and powder are mixed into the dough for this bread. Another example is the German pumpernickel, which uses commercial yeasts which enhanced with citric acid and lactic acid (Arici & Daglioglu, 2002) (Figure 7.1).

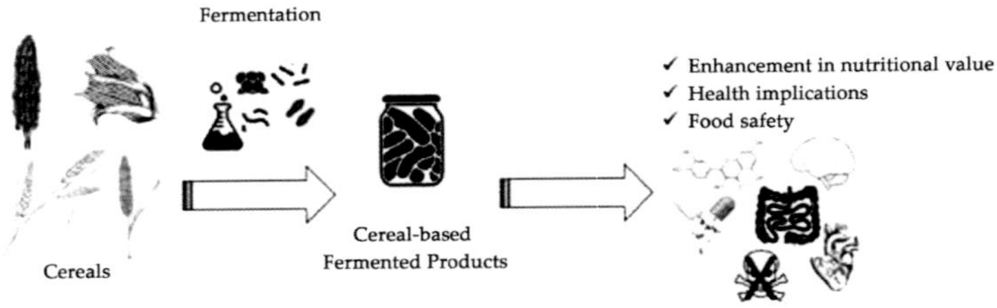

Figure 7.1. *Illustration of cereal-based fermented products.*

Source: Afaf Kamal-Eldin, Creative Commons License.

Various kinds of leavened breads are crafted from traditional cereals in Africa, such as the *Ethiopian injera* and the *Sudanese kisra*. The production of *Sudanese kisra* is made by sorghum and its natural microflora, which consists of various bacteria and molds. The bacteria present include *L. cellobiosus*, *Lactobacillus coprophilus*, *L. brevis*, *Klebsiella pneumoniae*, *Serratia ficaria* *Pediococcus pentosaceus*, *Enterobacter agglomerans*, and *Enterobacter sakazakii*. Additionally, molds such as *Fusarium* spp., *Penicillium* spp., *Aspergillus niger*, and *Alternaria* spp. are also involved in the fermentation process (Waters et al., 2015).

The fermentation process involves incorporating a portion of aged dough, known as kham'mar, into flour to create a dense dough or ajin. This mixture is then left to ferment for 12–24 hours, resulting in a decrease in pH from 5.5 to approximately 3.5. While glucose builds up as a bridge, the dough's maltose content steadily decreases. After being diluted, the ajin is deep-fried on a heated pan coated with oil using an oiled cloth. The resultant kisra bread is thin and bakes in a mere 60 seconds, emerging as a fragile and thin layer (Sharma et al., 2020).

In Ethiopia, the fluffy and pancakelike injera is a popular staple food. It is typically made from teff, a small grain native to the region. But it can also be made using sorghum, and different combinations of barley, wheat, or rice flour can be added. Similar to kisra, injera is created through the process of fermenting teff flour naturally or with the use of an aged sourdough starter (Lorenz et al., 1977).

After heating some of the fermented slurry to gelatinize the starch, it is blended with the raw portion and fermented for three to five hours. The process of fermenting injera leads to the breakdown of starch into free sugars, with fructose being the main component. The proportions of these sugars can vary depending on factors such as pH levels and the dynamics of the microbial population. In addition to being consumed in Eritrea, similar food items called canjeero or laxoox are enjoyed in Somalia, while Yemen has its version known as lahoh (Hachmeister & Fung, 1993) (Figure 7.2).

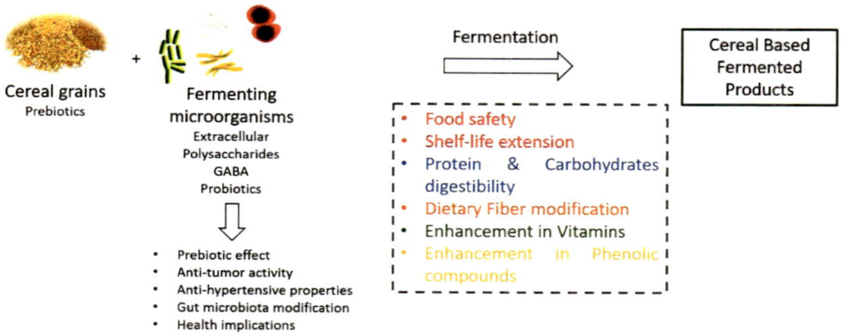

Figure 7.2. *Illustration of fermentation cereal products.*

Source: Eva Tejedora, Creative Commons License.

There is a diverse selection of porridges and fermented gruels found in Africa. These include edi, ogi, acidified thin ambali, ugi, thick asida, sankati, ugali, bogobe, and tuwo, which are all products of fermented maize and sorghum. Ogi is a smooth, creamy sour maize gruel is commonly consumed in Nigeria. It is a common weaning food for babies and a morning cereal for adults (Gobbetti et al., 2020).

Other African countries also have similar foods like kenkey or koko in mahewu and Ghana in South Africa. The fermentation process is controlled by lactic acid bacteria, specifically *Lb. planfreundii*, along with *Corynebacterium*, which breaks down the cornstarch, and yeasts, such as *Candida* and *Saccharomyces*, that play a role in flavor development (Copetti, 2019).

Furthermore, there is a wide production of fermented alcoholic and nonalcoholic beverages, such as borde, abrey, busa, ting, burkutu, and obushera. For example, borde is an Ethiopian beverage made by fermenting a combination of unmalted and malted flours derived from mixed or single cereals such as wheat, maize, barley, and finger millet (Guerra et al., 1999).

The borde production process includes various steps such as boiling, grinding, lactic acid fermentation, roasting, steam cooking, wet-milling, mashing, cooling, and wet-sieving. Malt is commonly utilized in the fermentation process of European and African beers, making up a significant portion of the substrate. The process of brewing merissa, the traditional sorghum beer in Sudan, is highly complex and requires high-end techniques. This process utilizes sorghum malt as an enzymatic preparation, rather than as a substrate, together with a caramelized sorghum product known as surij (Vogelmann et al., 2009).

To meet the enzyme demands for a combination of raw and gelatinized starch, the substrate is made by mixing partially cooked, overcooked, and fully cooked sections. In Sudan, a complex process involving about 40 steps is used to prepare a millet beer or clear sorghum called assaliya (or um-bilbil). This method is more complicated with the simpler procedures used for amgba or otika, which require less than 20 steps (Poutanen, 1997).

During the holy month of Ramadan, Sudanese people typically sip hilu-mur, a delicious fermented sorghum drink. When it comes to the process of fermentation reaction, one interesting method involves using sorghum, specifically the red feterita cultivar. This sorghum is milled and malted into flour, which is then added to previously cooked porridge made from unmalted sorghum. The ratio of flour to porridge can be either 1:1 or 1:2. This process will be initiated by the addition of malted flour, the process of amylolysis, which will cause the porridge to become liquid. The steaming porridge is gently poured from the cooking kettle into a prepared jar,will ferment for two to three days. Enzymes from the malt, a starting culture of *Lactobacillus*, and the natural flora of the sorghum are used in this transformation, which occurs in a warm part of the house (Vargas et al., 2012).

The primary microorganisms involved in hilu-mur fermentation were identified as *Leuconostoc*, *Lactobacillus*, *Gluconobacter*, *Acetobacter*, *Candida,* and the yeasts *Saccharomyces*. Additional ingredients are incorporated into the hill-mur mixture, which has a sour, sweetish, and brownish flavor, one night prior to baking. These ingredients include tamarind, karkade, cinnamon, ginger, ghurunjal, and dates slurry (Le Francois, 1988).

The bread made from the hill-mur batter is reddish, fluffy, and has a strong spice and malt aroma. After that, it is spread and baked on a hot pan using a thinner bread sheet, much like how kisra baking is done. The sheets are carefully folded and placed on a well-ventilated surface to ensure they are completely dry before being stored. To produce an appealing hill-mur beverage with a pleasing flavor that combines sweetness and sourness, the dehydrated sheets are immersed in water. After that, the beverage is filtered, cooled, and sugar-sweetened (Englyst et al., 2003).

7.2. TYPES OF FERMENTED LEGUME PRODUCTS AND THEIR MANUFACTURE

Learning Objectives:

- Understand the types and processes of fermented legume products.

Worldwide, many leguminous plants are fermented, though this process is not as common as toasting, boiling, and germination. One notable exception to this general rule is soybean. There is a diverse selection of fermented soybean products, primarily made from soybeans but also incorporating a variety of other ingredients such as cereals, chili, legumes, and spices (Reddy et al., 1983).

Soy sauce is widely recognized as a prominent outcome of soybean fermentation on a global scale. When it comes to the production of soy sauce, an interesting process takes place. Soybeans, either on their own or combined with cereal grains, undergo fermentation with the help of molds such as *Aspergillus oryzae* and *A. sojae*. This fermentation results in the creation of a substance called moromi. The moromi is then pressed to separate it into two components: a liquid, which we know as soy sauce and a solid cake commonly utilized as animal feed. Koji is typically used to refer to the fermentation inoculum (Adebo et al., 2022).

Some soy sauce brands are produced using acid-hydrolyzed soy protein, resulting in a longer shelf life and distinct aroma, flavor, and texture same as traditionally brewed soy sauces. Wheat, whether it is roasted or unroasted, is often combined with soybeans to create Japanese soy sauces. It can make up to 50% of the ingredients and adds a touch of sweetness to the sauce.

Did You Know?

Fermented legume products like tempeh originated in Indonesia, where they have been a dietary staple for centuries. Tempeh is made by fermenting cooked soybeans with a mold called Rhizopus oligosporus, resulting in a dense, nutty-flavored cake.

Tempeh, also known as tempe kedele, is a soy-based food and its roots in Indonesia and has gained popularity in neighboring Asian countries and beyond. This product is a soybean-grain creation that resembles a vegetarian burger patty. It has a rubbery, firm texture and a distinct earthy flavor that becomes more pronounced as it ages (Senanayake et al., 2023).

Tempeh, also known as tempeh kedelai, is created through a fermentation process involving dehulled and partially cooked soybeans. These soybeans are mixed with a starting culture containing *Rhizopus oligosporus* fungal spores. The fermentation takes place over 24–36 hours at a temperature of approximately 30°C–32°C. To create an ideal environment for the growth of the tempeh mold and prevent the growth of other microorganisms, it is common practice to add a mild acid, such as vinegar, to neutralize ammonia and lower the pH.

Throughout the process, the beans undergo fermentation and become connected with a mat of white or grayish mycelia. Various variations of tempeh can be created by making changes to the preparation process or by using different types of beans or combinations of whole-grain cereals and beans. For example, tempe mendoan is made by thinly slicing tempeh, coating it in batter, and frying it until it becomes crispy or partially cooked with a soft and tender texture (Parra et al., 2013).

Tempeh bacem is a flavorful variation of tempeh that combines a rich blend of spices and palm sugar. It is prepared by boiling the tempeh with these ingredients and then frying it briefly to elevate its taste. Tempe kering is created through the process of deep-frying raw tempeh that has been cut into small sticks. It is then combined with anchovies, fried peanuts, sugar, and spices (Curiel et al., 2015).

Tempehembus is crafted from soybean curd residue, while tempe bosok, or tempe busuk, is a unique variation of tempeh that offers a distinct and pungent flavor, often used sparingly as a seasoning. Tempe gódhóng is a type of tempeh that is delicately encased in banana leaves.

In addition to soybean, tempeh can be created using various legumes such as tempe gembus (made from okra), tempeh bengook (derived from *Mucana pruriens*), tempeh lamtoro (made from *Leucaena leucocephala*), tempeh koro (made from *Phaseolus lunatis*), and tempeh oncom (orange-colored tempeh made from peanut cake).

Tempeh bongkrèk, derived from coconut press cake, can sometimes be affected by the presence of Burkholderia gladioli, a bacterium that produces toxins known as toxoflavin and bongkrek acid. These two toxins' structures are shown in Figure 7.3. Bongkrek acid is extremely dangerous due to its ability to block the ATP/ADP translocation process, which hinders the release of ATP from the mitochondria and the supply of energy to the body. Due to its toxic nature, tempeh bongkrèk has been prohibited in Indonesia (Satusap et al., 2014).

Tofu, also known as soybean curd, has its roots in China and has since become popular in surrounding countries. Tofu is created by pressing of curds formed by coagulating proteins and oil found in soy milk, using a combination of salt or acid coagulants. Traditionally, calcium sulfate coagulants have been utilized in the production of Chinese-style tofu. This results in a tender and calcium-rich tofu, albeit with a slightly brittle texture.

Additionally, calcium chlorides and magnesium are utilized as coagulants in the production of tofu, resulting in a delightful texture both smooth and tender. Acidic coagulants, like vinegar and lemon juice, are commonly utilized in the coagulation process of soy milk to create tofu with varying flavors. Glucono-d-lactone, an acid coagulant, is used in the production of tofu to create a smooth and gelatinous texture (Chandra et al., 2016).

Coagulants are added to boiling soybean milk after being dissolved in water, and the mixture is stirred until it curdles into a soft gel of fresh tofu with different textures and microscopic characteristics, such as firm tofu, Chinese dry tofu, and soft "silken" tofu. Tofu that is still soft is always stored in water. Additional processing of fresh tofu can result in varieties that are fermented/ pickled, seasoned, fried, and dried.

Memo miso, or traditional miso, is a thick paste formed of salt that is fermented in Japan using the fungus kojikin to ferment rice, barley, or soybeans. Miso is used to make misoshiru soup, pickled veggies or meats, and sauces and spreads. It is also combined with dashi soup stock (Nagai & Tamang, 2010).

While mixed miso (awasemiso) is made from soybeans and other ingredients, white miso (shiromiso) and red miso (akamiso) are made from boiling and steamed soybeans, respectively. Maillard reactions cause the pigment of red miso throughout its comparatively longer fermentation process.

Other ingredients used to make misos include genmai from brown rice, kome from rice, tsubu from whole wheat and barley, mugi from barley, hadakamugi from rye, sobamugi from buckwheat, and taima from hemp seed. Misos are not only created from soybeans.

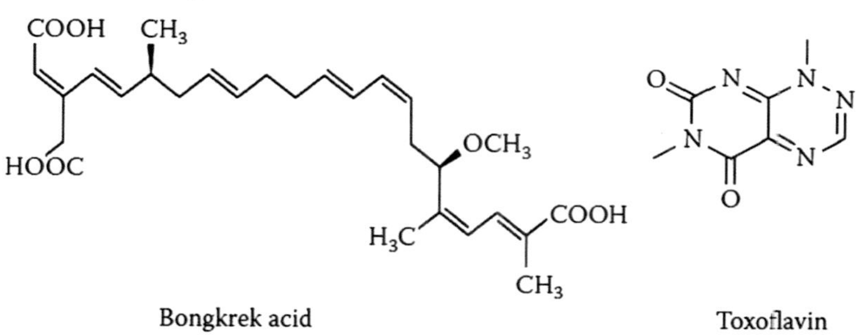

Figure 7.3. *Illustration of structures of bongkrek acid and toxoflavin.*

Source: Sam Jennings, Creative Commons License.

Processing factors such as grain type, color, quality, additional ingredients and additives, fermentation temperature, time (from five days to many years), and salt content all affect the final product's flavor and quality. New varieties of miso have been made recently in different nations with substitute components such as amaranth, adzuki beans, chickpeas, corn, and quinoa.

In Japan, soybeans undergo a fermentation process using Bacillus subtilis (commonly referred to as natto-kin) to create natto, a breakfast food known for its pungent aroma, distinct taste and slippery consistency. It is generally recommended to use smaller beans to ensure that the fermentation process can penetrate the center more effectively. The beans undergo a thorough washing then soaking in water and steaming. They are then mixed with the bacterium and subjected to fermentation at a temperature of 40°C for 24 hours (Arbab et al., 2023).

The natto is then aged at a low temperature to facilitate the development of bacteria spores and enzymatic peptidases that break down the soybean protein. Natto is thought to offer various health advantages, such as protecting the heart and preventing blood clots. This is linked to the presence of a serine protease enzyme called nattokinase, which can directly break down blood clots and inhibit the protein plasminogen activator inhibitor-1. Furthermore, natto contains a variety of beneficial compounds such as pyrazines, vitamin K_2, and phytoestrogenic soybean isoflavones. These compounds have been shown to have protective effects against a range of health conditions including cancer, Alzheimer's disease, osteoporosis, heart disease, and aging (Watson et al., 2017).

Ontjom is a type of fermented food is popular in Indonesia. It is made from peanut press cake, either on its own or mixed with cassava press cake, coconut press cake, and leftovers from making tofu or soybean milk. Additional ingredients are included, partially or entirely, to enhance texture and minimize expenses.

Fermentation is carried out by molds such as Rhizopus and Neurospora. If Rhizopus is the main mold, the fermented cake will have a gray look due to the abundance of blackish sporangia in it. Conversely, if there is a high concentration of Neurospora intermedia, the cake will turn pink because of the development of pink conidia (Garrido et al., 2021).

In India, for example, a variety of fermented products are derived from Bengal gram (= black chickpeas, Cicer arietinum) and black gram (Phaseolus mungo L.). These include vadai, papadam, bhallae, and warri (products made from black gram), as well as dosa and idli (products made from rice and black gram), khaman and dhokla (products made from Bengal gram). Occasionally, fermentation is applied to various legumes including velvet beans, melon seeds, winged beans, yellow chickpeas, and broad beans (Moio et al., 1993).

7.3. MICROORGANISMS USED IN FERMENTATION

Learning Objectives:

- Understand the role and use of microorganisms in fermentation.

Various microorganisms are playing their role in cereal fermentation, such as yeasts, bacteria, and molds. Numerous bacterial species, including *Leuconostoc*, *Pediococcus*, *Lactobacillus*, and *Lactococcus*, are essential for the formation of acids and tastes in these goods. The bacterial flora present in various fermented foods is influenced by multiple factors such as pH levels, salt concentration, water activity, temperature, and food matrix composition (Ross et al., 2002).

During LAB fermentations, the grains' own amylases produce sugars that serve as a source of energy for the bacteria. There are two main groups of lactic acid bacteria: homofermentative and heterofermentative. The homofermentative group includes *Streptococcus*, *Pediococcus*, *Lactococcus*, and some *Lactobacilli*. On the other hand, the heterofermentative group consists of *Weissella*, *Leuconostoc*, and some *Lactobacilli* (Cramer et al., 2005).

In the process of glucose fermentation, there are two types of microorganisms: homofermentative and heterofermentative. The principal byproduct of homofermentative microbes is lactic acid, whereas heterofermenters create a mixture of lactate, CO_2, and ethanol. The many kinds of microflora that might flourish in the fermentation medium are significantly impacted by the generation of these metabolites (Meynier et al., 1999).

Fun fact:

Natto, a popular Japanese fermented soybean dish, is known for its unique sticky texture and pungent aroma. It is commonly consumed with rice in the morning and is thought to offer various health advantages because of its rich protein and vitamin content.

Yeasts can thrive on glucose or maltose, breaking them down into carbon dioxide and alcohol, respectively. The production of beer through the process of alcoholic fermentation is primarily carried out by the brewer's yeast *Saccharomyces*, which converts maltose from malted grain into alcohol. Positive symbiotic relationships between lactic acid bacteria and yeasts are frequently seen in fermented foods (Caplice & Fitzgerald, 1999).

The process of making sourdough bread involves the combination of various microorganisms, including yeasts, lactic acid bacteria, and Torulopsis. These microorganisms interact with different cereal grains like rye and wheat to create a unique flavor and texture. Two different sorts of biochemical changes are catalyzed by yeast and lactobacilli: yeast uses carbon dioxide to leaven the dough, whereas lactobacilli employ lactic and acetic acids to acidify the dough. Fermented cereal products have a nice flavor that is mostly attributed to yeasts, specifically *Candida* spp. and *Saccharomyces cerevisiae*. Additionally, the distinctive sensory characteristics of these breads are attributed to hetero fermentative strains such as *Lb. brevis* and *Lb. san Francisco*.

Majority of the microbes responsible for the fermentation of tempe kedele are fungi in the Rhizopus genus, specifically *R. oligosporus*, which yields the end products, which are compact cakes knitted from mycelia. Other two Aspergillus species and Mucor species aremicroorganisms that are less common in tempehs (Xu et al., 2021).

A species of fungus belonging to the Aspergillus genus called koji is essential to the fermentation process of some soybean products. Due to their strong proteolytic properties, Aspergillus oryzae and *Aspergillus sojae* strains are the main species engaged in these fermentations. Furthermore, tempeh and koji are enriched with *Lactobacillus* spp. which contribute to the lactic acid production, enhancing the product's acidity. Additionally, *Bacillus* spp. play a role in generating distinct odors and ammonia (Shay et al., 1987).

7.4. CHEMICAL CHANGES IN FERMENTED FOODS

Learning Objectives:

- Understand the chemical changes induced by pre-fermentation treatments.
- Identify the changes in carbohydrates, proteins, lipids, minerals, and vitamins during fermentation.
- Recognize the role of volatile compounds in fermented food flavors

7.4.1. Chemical Changes Induced by Pre-Fermentation Processes

Depending on the type of raw material and the intended final product, different pre-fermentation procedures are available for cereal and legume products. Some of the processes involved are washing, steeping, drying, fractionation, milling, germination, sieving, and cooking. Cereals and pulses that have been malted or sprouted are frequently utilized as components in a variety of fermented dishes. For example, starch gelatinizes when cereal flour is boiled with porridge at a temperature between 80°C and 95°C prior to fermentation. This leads to a final product that is thick and has a high viscosity, which in turn enhances the fermentation process with the help of malt enzymes (Moulin et al., 1999).

Adding malted flour serves two purposes: it enhances the activity of amylase enzymes and helps to decrease the cooked flour viscosity. During the malting process of cereal grains, a variety of beneficial transformations take place as complex compounds are broken down into simpler ones and essential constituents are formed which are easy to digest. (Wang et al., 2016).

The cooked product becomes less thick and is ideal for fermenting with lactic acid bacteria like Bifidobacteria. Furthermore, sprouting enhances the taste, ease of digestion, and nutritional content of these food items. In addition to beverages, malted cereals are commonly incorporated into recipes for breakfast cereals, biscuits, breads, other bakery products, and confectionaries (Alti-Palacios et al., 2023).

Preparation methods before fermentation can sometimes result in a decrease in the nutritional content of the final products, which is particularly significant for staple foods like ogi. Efforts to enhance the nutritional value of these staple foods have focused on strengthening them with legumes to address the amino acid deficiencies. Several methods have been explored to enhance the nutritional value of ogi, such as incorporating legumes, both unfermented and fermented, and adding pawpaw slurry as a substitute at different levels (Parley et al., 2001).

7.4.2. Chemical Changes Induced by Fermentation

7.4.2.1. Changes in Carbohydrates

Digestible carbohydrate polymers, such as starch and various forms of fiber, can be broken down into mono- and oligosaccharides by fermenting bacteria. Enzymes that hydrolyze carbohydrates, including glucosidases, fructofuranosidases, galactanases, and amylases, are known to be present in Bacillus species. These enzymes hydrolyze carbs into sugars, which the microbiota uses for food and convert into other compounds like pyruvate and ethanol in part. Gases (carbon dioxide, methane, hydrogen sulfide, and hydrogen), short-chain fatty acids (SCFA, propionic, primarily acetic, and butyric acids), and lactic acid in the case of *Lactobacillus* spp. are the results of this fermentation (Evans & Reid, 1988).

7.4.2.2. Changes in Proteins

The fermented raw cereal and grain materials undergo many types of changes as a result of the endogenous enzymes and the enzymatic activity of fermentative yeasts and bacteria. The sourdough fermentation, caused by changes in pH, can activate certain enzymes in grains, such as cereal proteases that break down endogenous prolamin. Consequently, this can facilitate the decomposition of gliadins, glutenins, globulins, and secalins, as well as the disintegration and dissolution of gluten networks (Lumry & Eyring, 1954).

Practice Problem

Discuss the changes in proteins that occur during the fermentation of legume products and their implications for nutritional quality.

Solutions to Practice Problem

Proteolysis is the process by which bacteria produce enzymes that break down large proteins into simpler peptides and amino acids during the fermentation of legume products. The process by which proteins are broken down by enzymes increases the bioavailability and digestibility of amino acids, facilitating the body's easier absorption. Additionally, fermentation can lead to the synthesis of bioactive peptides with potential health benefits, such as antioxidant and antimicrobial properties. For example, in the fermentation of soybeans to produce tempeh, proteolysis results in the formation of peptides with angiotensin-converting enzyme (ACE) inhibitory activity, which may help regulate blood pressure. Overall, the nutritional value and health-promoting qualities of legume products are enhanced by the fermentation-induced modifications in proteins.

Cereal proteins are hydrolyzed by the different proteolytic processes brought on by fermentation, results in the production of free amino acids, which serve as taste precursors. Yeasts utilize free amino acids for their metabolic processes, while bacterial fermentation will lead to an elevation in the concentration of free amino acids in dough. Thus, the amount of free amino acids in the dough is dependent on the fermentation culture's makeup, temperature, and time, as well as the presence of substrates (Tama & Sanejouand, 2001).

These conditions have varying effects on the individual amino acids. For example, the pH level plays a major role in releasing different amino acids. Proline is more likely to be released when the pH is above 5.5, whereas phenylalanine, leucine, and cysteine are released at lower pH levels. Ornithine, a key component in enhancing the flavor of bread, is exclusively present in dough that undergoes fermentation with *Lactobacillus pontis*. Brans and whole-grain flours exhibit higher proteolytic activities in comparison to white flour (Rasper & Kauzmann, 1962) (Figure 7.4).

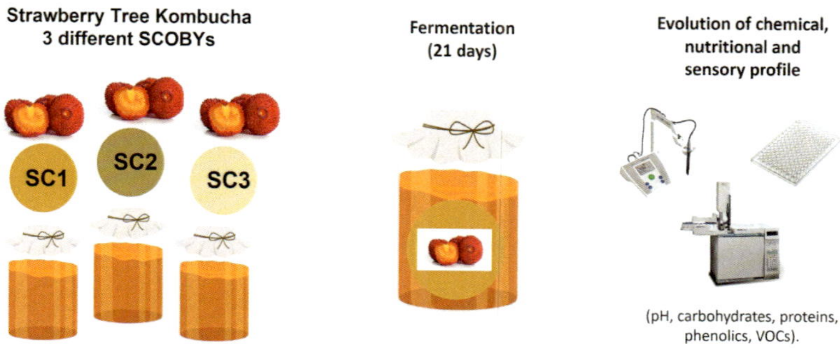

Figure 7.4. *Illustration of chemical changes in fermentation.*

Source: Diego Morals, Creative Commons License.

7.4.2.3. Changes in Lipids

During the fermentation process, certain reactions can occur that result in the release of free fatty acids, including oleic, linoleic, and linolenic acids. These acids have the potential to negatively impact the sensory and nutritional characteristics of the final products. However, although oils account for a large fraction of the legumes used in food fermentations, only small amounts of lipase activity and minimal breakdown of fats were observed in some examples, including the fermentation of melon seeds, *Pentaclethra macrophylla* during ugba production, and *Parkia biglobosa* during dawadawa production (Chaijan et al., 2006).

There have been reports of aldehydes resulting from the oxidation of lipids, which can cause off-flavors. However, there have been reports of the positive impact of lipase on the development of distinct aroma and flavor. Despite attempts to remove lipid-soluble odorants in fermented products from melon seed, locust bean, and soybean using organic solvents, these efforts proved unsuccessful, highlighting the complicated nature of this problem (Kattner & Krause, 1987).

7.4.2.4. Changes in Vitamins and Minerals

Minerals are not naturally incorporated into the fermented foods. Nevertheless, fermentation helps to maintain the ideal pH for cereal phytases that facilitate the enzymatic breakdown of phytates. Phytates found in cereals and legumes have remarkable ability to form insoluble complexes with essential metal ions like zinc, iron, calcium, and magnesium, hindering their absorption in the body (Ren et al., 2024).

In Chapter 6, a thorough discussion is provided on how fermentative microorganisms can improve the levels and availability of various vitamins. These include B2 (riboflavin), B1 (thiamine), B7 (biotin), B6 (pyridoxine, pyridoxamine, and pyridoxal), B5 (pantothenic acid), B12 (cyanocobalamin), folate, and K (Grzybek et al., 2002).

7.4.2.5. Volatile Compounds

Bacterial fermentation generally yields acids (such as butyric, propionic, and lactic acids), whereas yeast fermentation yields alcohol and more volatile substances. Bacteria produce volatiles according to their strain: yeast produces ethyl acetate and iso-alcohols like 3-methylbutanol, 2-methyl-1-propanol, 2,3-methyl-1-butanol, and 2-phenylethanol; heterofermentative lactobacilli produces ethyl acetate, alcohols, and aldehydes; and homo fermentative lactobacilli produces acetaldehyde, high levels of diacetyl, and hexanal. Important flavoring chemicals generated during sourdough fermentation include acetic acid, butanoic acid, phenylacetic acid, 2 and 3-methylbutanoic acid, and pentanoic acid (Yang et al., 2013).

SUMMARY

- The chapter addresses the global practice of fermenting legumes and cereal grains, emphasizing the advantages of fermentation, including improved flavor, greater nutritional value, and preservation.
- It describes the various fermentation processes involved, such as lactic acid, acetic acid, and alcoholic fermentation, as well as the wide range of goods they produce, including beverages, gruels, bread, and soy sauce made from cereals and miso, tempeh, tofu, and other legumes.
- Variety of microorganisms, including bacteria, molds, and yeasts, are essential for fermentation and provide flavor and texture.
- The breakdown of proteins, lipids, and carbohydrates as well as the enrichment of minerals and vitamins during fermentation are highlighted as chemical changes that occur during fermentation and lead to the production of distinctive and nutrient-dense fermented meals in many cultures.

REVIEW QUESTIONS

1. What are the main types of fermented cereal products discussed in the chapter, and how are they manufactured?
2. Name two fermented legume products covered in the chapter and briefly describe their manufacturing processes.
3. What role do microorganisms play in fermentation, and how do they differ in their contributions to cereal and legume fermentation?
4. Explain the chemical changes induced by fermentation in fermented foods, focusing on carbohydrates, proteins, and lipids.
5. How do pre-fermentation processes influence the subsequent fermentation and final characteristics of fermented products?
6. Discuss the significance of volatile compounds in fermented foods and their impact on aroma and flavor profiles.
7. What are the primary factors influencing the choice of microorganisms for fermentation in cereal and legume products?
8. How do fermented cereal and legume products contribute to dietary diversity and nutritional value?

MULTIPLE CHOICE QUESTIONS

1. What type of microorganisms are commonly used in fermentation?

a. Bacteria, yeast, and molds

b. Insects

c. Birds

d. Reptiles

2. Which of the following is a fermented legume product?

a. Yogurt

b. Tempeh

c. Milk

d. Butter

3. What chemical changes occur during fermentation?

a. No changes

b. Changes only in carbohydrates

c. Changes in carbohydrates, proteins, and lipids

d. Changes only in proteins

4. What role do volatile compounds play in fermented foods?

a. They have no role

b. They contribute to aroma and flavor

c. They make the food toxic

d. They reduce the shelf life

5. Which pre-fermentation process influences subsequent fermentation processes?

a. Heating

b. Cooling

c. Washing

d. Grinding

6. What chemical changes are induced by fermentation?

a. Changes in color only

b. Changes in texture only

c. Changes in flavor, aroma, and nutritional content

d. Changes in appearance only

Answers to Multiple Questions

1. (a); 2. (b); 3. (c); 4. (b); 5. (d) 6. (c)

REFERENCES

1. Abou-Zeid, N. A. (2016). Review of Egyptian cereal-based fermented product (Kishk). *International Journal of Agriculture Innovations and Research, 4*(4), 600–609.
2. Achi, O. K., & Ukwuru, M. (2015). Cereal-based fermented foods of Africa as functional foods. *International Journal of Microbiology and Application, 2*(4), 71–83.
3. Adebo, J. A., Njobeh, P. B., Gbashi, S., Oyedeji, A. B., Ogundele, O. M., Oyeyinka, S. A., & Adebo, O. A. (2022). Fermentation of cereals and legumes: Impact on nutritional constituents and nutrient bioavailability. *Fermentation, 8*(2), 63.
4. Adebo, O. A., Oyeyinka, S. A., Adebiyi, J. A., Feng, X., Wilkin, J. D., Kewuyemi, Y. O., & Tugizimana, F. (2021). Application of gas chromatography–mass spectrometry (GC-MS) -based metabolomics for the study of fermented cereal and legume foods: A review. *International Journal of Food Science & Technology, 56*(4), 1514–1534.
5. Alti-Palacios, L., Martínez, J., Teixeira, J. A., Câmara, J. S., & Perestrelo, R. (2023). Influence of cold pre-fermentation maceration on the volatilomic pattern and aroma of white wines. *Foods, 12*(6), 1135.
6. Arbab Sakandar, H., Chen, Y., Peng, C., Chen, X., Imran, M., & Zhang, H. (2023). Impact of fermentation on antinutritional factors and protein degradation of legume seeds: A review. *Food Reviews International, 39*(3), 1227–1249.
7. Arici, M., & Daglioglu, O. (2002). Boza: A lactic acid fermented cereal beverage as a traditional Turkish food. *Food Reviews International, 18*(1), 39–48.
8. Ashaolu, T. J., & Reale, A. (2020). A holistic review on Euro-Asian lactic acid bacteria fermented cereals and vegetables. *Microorganisms, 8*(8), 1176.
9. Blandino, A., Al-Aseeri, M. E., Pandiella, S. S., Cantero, D., & Webb, C. (2003). Cereal-based fermented foods and beverages. *Food Research International, 36*(6), 527–543.
10. Caplice, E., & Fitzgerald, G. F. (1999). Food fermentations: Role of microorganisms in food production and preservation. *International Journal of Food Microbiology, 50*(1–2), 131–149.

11. Chaijan, M., Benjakul, S., Visessanguan, W., & Faustman, C. (2006). Changes of lipids in sardine (*Sardinella gibbosa*) muscle during iced storage. *Food Chemistry, 99*(1), 83–91.
12. Chandra-Hioe, M. V., Wong, C. H., & Arcot, J. (2016). The potential use of fermented chickpea and faba bean flour as food ingredients. *Plant Foods for Human Nutrition, 71*, 90–95.
13. Chavan, J. K., Kadam, S. S., & Beuchat, L. R. (1989). Nutritional improvement of cereals by fermentation. *Critical Reviews in Food Science & Nutrition, 28*(5), 349–400.
14. Copetti, M. V. (2019). Yeasts and molds in fermented food production: An ancient bioprocess. *Current Opinion in Food Science, 25*(3), 57–61.
15. Cramer, A. C. J., Mattinson, D. S., Fellman, J. K., & Baik, B. K. (2005). Analysis of volatile compounds from various types of barley cultivars. *Journal of Agricultural and Food Chemistry, 53*(19), 7526–7531.
16. Curiel, J. A., Coda, R., Centomani, I., Summo, C., Gobbetti, M., & Rizzello, C. G. (2015). Exploitation of the nutritional and functional characteristics of traditional Italian legumes: The potential of sourdough fermentation. *International Journal of Food Microbiology, 196*(1), 51–61.
17. Englyst, K. N., Vinoy, S., Englyst, H. N., & Lang, V. (2003). Glycaemic index of cereal products explained by their content of rapidly and slowly available glucose. *British Journal of Nutrition, 89*(3), 329–339.
18. Evans, R. Y., & Reid, M. S. (1988). Changes in carbohydrates and osmotic potential. *Journal of the American Society for Horticultural Science, 113*(6), 884–888.
19. Garrido-Galand, S., Asensio-Grau, A., Calvo-Lerma, J., Heredia, A., & Andrés, A. (2021). The potential of fermentation on nutritional and technological improvement of cereal and legume flours: A review. *Food Research International, 145*, 110398.
20. Gobbetti, M., De Angelis, M., Di Cagno, R., Polo, A., & Rizzello, C. G. (2020). The sourdough fermentation is the powerful process to exploit the potential of legumes, pseudo-cereals and milling by-products in baking industry. *Critical Reviews in Food Science and Nutrition, 60*(13), 2158–2173.
21. Grzybek, A., Klosiewicz-Latoszek, L., & Targosz, U. (2002). Changes in the intake of vitamins and minerals by men and women with hyperlipidemia and overweight during dietetic treatment. *European Journal of Clinical Nutrition, 56*(12), 1162–1168.
22. Guerra-Hernandez, E., Corzo, N., & Garcia-Villanova, B. (1999).

Maillard reaction evaluation by furosine determination during infant cereal processing. *Journal of Cereal Science, 29*(2), 171–176.

23. Hachmeister, K. A., & Fung, D. Y. (1993). Tempeh: A mold-modified indigenous fermented food made from soybeans and/or cereal grains. *Critical Reviews in Microbiology, 19*(3), 137–188.
24. Kattner, G., & Krause, M. (1987). Changes in lipids during the development of Calanus finmarchicus sl from copepodid I to adult. *Marine Biology, 96*, 511–518.
25. Le Francois, P. (1988). Phytic acid and zinc contents of cereal products: Relation to the manufacturing process: 1. Phytic acid. *Journal of Food Composition and Analysis, 1*(2), 139–145.
26. Lorenz, K., Lee, V. A., & Jackel, S. S. (1977). The nutritional and physiological impact of cereal products in human nutrition. *Critical Reviews in Food Science & Nutrition, 8*(4), 383–456.
27. Lumry, R., & Eyring, H. (1954). Conformation changes of proteins. *The Journal of Physical Chemistry, 58*(2), 110–120.
28. Meynier, A., Novelli, E., Chizzolini, R., Zanardi, E., & Gandemer, G. (1999). Volatile compounds of commercial Milano salami. *Meat Science, 51*(2), 175–183.
29. Moio, L., Dekimpe, J., Etievant, P., & Addeo, F. (1993). Neutral volatile compounds in the raw milks from different species. *Journal of Dairy Research, 60*(2), 199–213.
30. Moulin, A. M., O'Shea, S. J., Badley, R. A., Doyle, P., & Welland, M. E. (1999). Measuring surface-induced conformational changes in proteins. *Langmuir, 15*(26), 8776–8779.
31. Nagai, T., & Tamang, J. P. (2010). Fermented legumes: Soybean and non-soybean products. In J. Tamang & K. Kailasapathy (Eds.), *Fermented Foods and Beverages of the World* (pp. 191–224).
32. Parley, A., Vanhanen, L., & Heatherbell, D. (2001). Effects of pre-fermentation enzyme maceration on extraction and color stability in Pinot Noir wine. *Australian Journal of Grape and Wine Research, 7*(3), 146–152.
33. Parra, K., Ferrer, M., Piñero, M., Barboza, Y., & Medina, L. M. (2013). Use of *Lactobacillus acidophilus* and *Lactobacillus casei* for a potential probiotic legume-based fermented product using pigeon pea (*Cajanus cajan*). *Journal of Food Protection, 76*(2), 265–271.
34. Poutanen, K. (1997). Enzymes: An important tool in the improvement of the quality of cereal foods. *Trends in Food Science & Technology, 8*(9), 300–306.

35. Rasper, J., & Kauzmann, W. (1962). Volume changes in protein reactions. I. Ionization reactions of proteins. *Journal of the American Chemical Society, 84*(10), 1771–1777.
36. Reddy, N. R., Pierson, M. D., Sathe, S. K., Salunkhe, D. K., & Beuchat, L. R. (1983). Legume-based fermented foods: Their preparation and nutritional quality. *CRC Critical Reviews in Food Science and Nutrition, 17*(4), 335–370.
37. Reddy, N. R., Sathe, S. K., & Salunkhe, D. K. (1982). Phytates in legumes and cereals. *Advances in Food Research, 28*, 1–92.
38. Ren, Q., Li, K., Li, J., Pan, J., Liu, Y., Chen, Y., & Xie, Q. (2024). Longitudinal changes in human milk minerals and vitamins in the Chinese population: A scoping review. *Nutrients, 16*(11), 1710.
39. Ross, R. P., Morgan, S., & Hill, C. (2002). Preservation and fermentation: Past, present and future. *International Journal of Food Microbiology, 79*(1–2), 3–16.
40. Satusap, P., Chavasit, V., Kriengsinyos, W., & Judprasong, K. (2014). Development of cereal and legume-based food products for the elderly. *SpringerPlus, 3*(1), 1–8.
41. Senanayake, D., Torley, P. J., Chandrapala, J., & Terefe, N. S. (2023). Microbial fermentation for improving the sensory, nutritional and functional attributes of legumes. *Fermentation, 9*(7), 635.
42. Sharma, R., Garg, P., Kumar, P., Bhatia, S. K., & Kulshrestha, S. (2020). Microbial fermentation and its role in quality improvement of fermented foods. *Fermentation, 6*(4), 106.
43. Shay, L. K., Hunt, H. R., & Wegner, G. H. (1987). High-productivity fermentation process for cultivating industrial microorganisms. *Journal of Industrial Microbiology, 2*(2), 79–85.
44. Tama, F., & Sanejouand, Y. H. (2001). Conformational change of proteins arising from normal mode calculations. *Protein Engineering, 14*(1), 1–6.
45. Vargas, F., González, Z., Sánchez, R., Jiménez, L., & Rodríguez, A. (2012). Cellulosic pulps of cereal straws as raw material for the manufacture of ecological packaging. *Bioresources, 7*(3), 4161–4170.
46. Vogel, R. F., Pavlovic, M., Ehrmann, M. A., Wiezer, A., Liesegang, H., Offschanka, S., & Liebl, W. (2011). Genomic analysis reveals *Lactobacillus sanfranciscensis* as stable element in traditional sourdoughs. In *Microbial Cell Factories* (Vol. 10, pp. 1–11).
47. Vogelmann, S. A., Seitter, M., Singer, U., Brandt, M. J., & Hertel, C. (2009). Adaptability of lactic acid bacteria and yeasts to sourdoughs

prepared from cereals, pseudocereals and cassava and use of competitive strains as starters. *International Journal of Food Microbiology, 130*(3), 205–212.

48. Wang, J., Huo, S., Zhang, Y., Liu, Y., & Fan, W. (2016). Effect of different pre-fermentation treatments on polyphenols, color, and volatile compounds of three wine varieties. *Food Science and Biotechnology, 25*, 735–743.
49. Waters, D. M., Mauch, A., Coffey, A., Arendt, E. K., & Zannini, E. (2015). Lactic acid bacteria as a cell factory for the delivery of functional biomolecules and ingredients in cereal-based beverages: A review. *Critical Reviews in Food Science and Nutrition, 55*(4), 503–520.
50. Watson, C. A., Reckling, M., Preissel, S., Bachinger, J., Bergkvist, G., Kuhlman, T., & Stoddard, F. L. (2017). Grain legume production and use in European agricultural systems. *Advances in Agronomy, 144*, 235–303.
51. Xu, Y., Zang, J., Regenstein, J. M., & Xia, W. (2021). Technological roles of microorganisms in fish fermentation: A review. *Critical Reviews in Food Science and Nutrition, 61*(6), 1000–1012.
52. Yang, Z., Baldermann, S., & Watanabe, N. (2013). Recent studies of the volatile compounds in tea. *Food Research International, 53*(2), 585–599.

CHAPTER 8

FERMENTED CEREAL AND LEGUME PRODUCTS

LEARNING OBJECTIVES

At the end of this chapter, readers will be able to:

1. Understand the role of lactic acid fermentation in enhancing food safety.
2. Identify key factors influencing the safety and quality of fermented foods.
3. Know the nutritional benefits associated with fermented foods.
4. Evaluate the potential hazards of fermentation in food processing.
5. Explain the application of HACCP principles to improve safety in fermented food production.

INTRODUCTORY EXAMPLE

A multinational food-products corporation "Danone," based in France, showcases the practical uses of fermentation. This company employs fermentation for various products, including traditional dairy items like yogurt and plant-based alternatives like soy and almond milk. By implementing controlled fermentation procedures, Danone can improve the nutritional content and extend its products' shelf life while maintaining strict food safety standards. This example demonstrates the potential for large-scale food companies to utilize fermentation technology to cater to consumer preferences for a wide range of nutritious and environmentally friendly food options.

UNIT INTRODUCTION

Fermentation is well known and utilized for centuries as a method of preserving food. Throughout history, fermentation has undergone significant advancements and adaptations, resulting in a wide range of food products now produced using this technique. These products are found in various settings, including large commercial enterprises, households, and small-scale food industries, and in both industrialized and developing nations. Fermented foods are widely consumed in different cultures across the globe. Fermentation is widely utilized in certain regions, particularly in African countries, to enhance the nutritional value of complementary foods for young kids and infants (Levin et al., 2004).

Recent developments in the field of food technology and science have resulted in the emergence of various innovative food technologies. Despite its long history, fermentation continues to be a crucial method in food processing. Fermentation is associated with numerous advantages. Fermentation has various benefits such as extending the shelf life of food, making it easier to digest, adding nutritional value, and enhancing its taste and flavor (Ye et al., 2021). Additionally, this technology is cost-effective, making it accessible to a wide range of populations. In addition, fermentation can improve food safety by effectively managing a wide range of pathogens in various food products. Therefore, it plays a crucial role in improving human nutrition, especially in regions with limited resources that face significant challenges in ensuring food security (Waché et al., 2018).

In general, fermented foods, particularly those produced in controlled settings, have a proven thetrack record of safety and are rarely linked to disease outbreaks. It is traditional along with artisanal productions often rely on the expertise and wisdom passed down through generations of food manufacturers and households, rather than academic research and formal applications of fermentation. This approach has a significant drawback. Depending on various factors such as the process, raw material, ingredients, and environmental conditions, there is a possibility of producing unsafe items (Verbelen et al., 2006).

This chapter explores the importance of fermentation in public health and food safety. The emphasis is placed on tackling difficulties in poor countries, where the risks of food contamination and disease are increased, and where the importance of fermentation in food processing and storage is especially notable. This chapter explores the potential advantages and disadvantages of fermentation in human nutrition. The study focuses on the fermentation process of food supplements and investigates methods to advance the technology and improve the safety and nutritional quality of the final products. Furthermore, it demonstrating the application of the Hazard Analysis and Critical Control Point (HACCP) approach to fermented foods using many illustrations (Gernaey et al., 2010).

8.1. BENEFITS OF LACTIC ACID FERMENTATION

Learning Objectives:

- Understand the benefits of lactic acid fermentation for inhibiting pathogens and enhancing food safety.
- Recognize the nutritional and health advantages of lactic acid fermentation in food products.

Different processes of fermentation exist, based on the organism employed, such as molds, yeasts, and bacteria. The processes of fermentation caused by lactic acid bacteria (LAB) is currently a topic of great interest in food safety and public health. Various forms of fermentation play a crucial role in safeguarding public health by potentially enhancing food security and addressing malnutrition concerns. This chapter primarily explores lactic fermentation (Gilliland, 1990).

Extensive research has conclusively demonstrated how lactic acid fermentation has a substantial effect on the development, survival, and toxin generation of numerous detrimental bacteria. The primary cause of this inhibitory impact is mainly related to the fast proliferation and acidogenesis by LAB, leading to a reduction in pH. Additionally, LAB also generates other antimicrobial substances like ethanol, bacteriocins, hydrogen peroxide, and diacetyl, further contributing to its effectiveness against pathogenic bacteria and spoilage organisms. The inhibition of pathogens is influenced by various factors, including the type of organism, temperature, acid production, and food properties such as buffering capacity. In the case of cereals and vegetable products with low buffering capacity, a successful lactic acid fermentation can lower the pH to 4 or below. This acidic environment inhibits growth of bacterial pathogens and leads to the death of many bacteria. The significance of the topic of fermentation, as discussed in the subsequent paragraphs, cannot be overstated in terms of its impact on public health. This is especially true in developing nations, where the challenges of ensuring safe food storage through refrigeration or heating are exacerbated by socioeconomic limitations (Lee, 1997) (Figure 8.1).

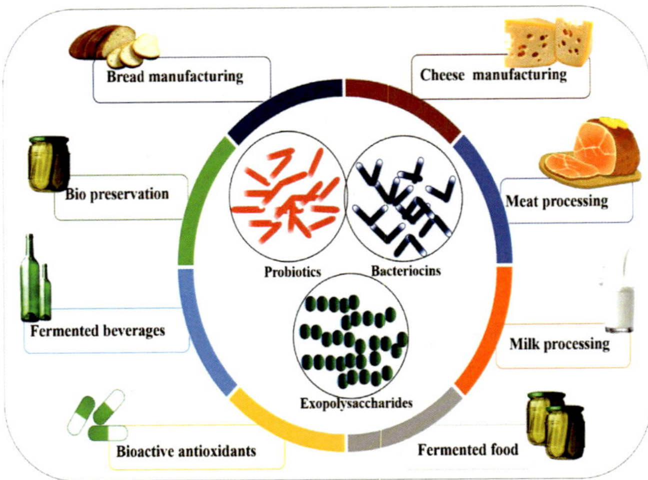

Figure 8.1. *Illustration of lactic acid bacterial fermentation in food.*

Source: Tirath Raj, Creative Commons Lice

Prior art suggests that lactic acid fermentation to the decrease of specific toxins found in plant foods, such as cyanogenic glycosides in cassava. Cassava is widely consumed in Africa and has been linked to various health issues because of the existence of certain compounds: Microbial processes that occur during fermentation play a crucial role in the process of detoxification by softening plant tissues, Furthermore, the breakdown of cyanogenic glycosides is facilitated by plant enzymes that are naturally present (Mathur et al., 2020).

Fermentation has also been associated with various nutritional benefits. Fermentation processes can improve the digestibility of carbohydrates by breaking down oligosaccharides and dietary fiber, is commonly found in plant-based foods. Combining amylase-rich flour (ARF) alongside a lactic acid starter culture may improve the nutritional value of starchy foods while maintaining a semiliquid texture. The fermented food is especially intriguing as a complementary food for babies and young children. Fermentation creates an ideal pH environment for the efficient functioning of phytase or the breakdown of phytate. The breakdown of phytate significantly enhances the solubility of iron, resulting in a substantial increase in its bioavailability. One interesting effect of lactic fermentation is its ability to reduce the tannin contents of cereals. This can have a beneficial effect on the absorption of minerals and the digestibility of proteins in cereals (Chelule et al., 2010).

In addition to its impact on pathogenic bacteria and toxic compounds, fermentation has been found to enhance food safety and quality. Certain strains of LAB have also been associated with beneficial health effects, including protection against Escherichia coli and various other pathogens, as well as hypocholesterolemic and anti-carcinogenic effects. Several organisms with purported health benefits are currently utilized in the manufacturing of food and are accessible in numerous countries. Two fundamental questions arise when considering the use of probiotics: (i) Are the organisms utilized considered safe? (ii) it is important to assess the validity of the health claims being made (Rodríguez et al., 2021).

Multiple research have shown specific health advantages associated with lactose intolerance and gastrointestinal issues. At now, there is no conclusive elucidation for the molecular pathways implicated. Moreover, other claims lack substantiating proof and necessitate further investigation (Fadlillah et al., 2021).

Innovations in modifying genes have greatly increased the potential of fermentation technology. Genetic modification might be applied to the substrate, including plants, or even to the microbes used in fermentation. Fermentation might be employed to generate digestive enzymes that exhibit superior features and enhanced quality. The utilization of fermentation can yield substantial advantages in enhancing the method of fermentation and enhancing the level of quality and safety of the end product (Kang, 2013). Various methods of genetic modification may be employed to optimize the process of fermentation and elevate the overall quality and security of the end product. As an example, starter cultures can be enhanced in terms of their growth rate, ability to produce inability to metabolize organic acids, production of flavor compounds, resistance to bacteriophage, lactic acid, tolerance to salt, and more. The application of genetic modification is the reduction or elimination of hazards in raw materials. Scientists have created a genetically engineered type of cassava that has reduced amounts of cyanogenic glycosides, and hence increasing its safety for consumption (Mazahreh & Ershidat, 2009).

To optimize the management of fermentation processes and enhance the safety and quality of resulting products, two essential requirements must be met: the utilization of genetic changes: (i) a reliable approach to evaluating the security and nutritional consequences of these goods; and (ii) consumer acceptability. Several agencies, such as FAO and WHO, developed ways to assess the safety of food. Furthermore, there are intentions to do further research in the future. To ensure consumer acceptance, it is crucial to consider the wider implications of genetic modification. This involves evaluating the potential hazards and advantages for customers, the environment, and the socioeconomic aspects in both emerging and developed nations. It is important to evaluate the necessity of such technology carefully. Effective communication with consumers is crucial when considering the potential use of genetic modification technology to enhance fermentation processes (Khubber et al., 2022).

Practice Problem

Explain why lactic acid fermentation is considered beneficial for food preservation. Provide two specific examples of foods where lactic acid fermentation is utilized and describe how it improves their shelf life.

Solutions to Practice Problem

Lactic acid fermentation is beneficial for food preservation because it lowers the pH of the food, creating an acidic environment that inhibits the growth of spoilage and pathogenic microorganisms. Examples include:

- **Sauerkraut:** Lactic acid fermentation of cabbage extends its shelf life by several months due to the acidic environment created by lactobacilli.
- **Yogurt:** Fermentation of milk by lactic acid bacteria not only preserves it but also enhances its nutritional value through the production of vitamins and enzymes.

8.2. PITFALLS OF FERMENTATION

Learning Objectives:

- Understand the pitfalls and risks associated with fermentation technologies in food production.

Despite its apparent simplicity, fermentation is a highly intricate and delicate process. Ensuring the safety as well as the quality of the result requires meticulous control over various factors. These factors are:

- The quality and security of raw materials, especially when fresh out of contamination, are crucial factors to consider.

- The importance of maintaining high levels of environmental sanitation and hygiene.
- Ensuring the efficacy and safety of starter cultures.
- Ensuring the security of metabolites.
- Factors such as processing conditions and the level of acidity attained play a crucial role in the fermentation process (Isenring et al., 2023).

In certain scenarios, especially when considering the utilization of fermentation in small-scale industries or domestic settings, it can be challenging toregulate these factors effectively. At times, fermented products were associated with infections or intoxications. In certain situations, inadequate fermentation has been a major factor of foodborne intoxication. As an example, China experienced a significant outbreak of botulism in 1989, with 2,861 reported cases and 421 fatalities. Homemade fermented bean goods were found to be the main foods involved. Botulism outbreaks are a serious concern for the native Alaskan Inuit populations in North America, often linked to the consumption of fermented fish and salmon eggs (Adams, 2014). Fermented Inuit items like salmon eggs, seal flippers, and white whale meat do not go through a sufficient pH decrease to prevent toxigenesis due to the absence of fermentable carbohydrates. Additionally, cheeses from various categories have been linked to outbreaks of different kinds of food-borne diseases. Unfortunately, the current foodborne disease tracking system has some limitations, especially when it comes to monitoring outbreaks caused by household food preparation. As a result, these outbreaks often go unnoticed and unreported (van de Sande, 1997).

It is important of maintaining a high level of hygiene in the raw materials and minimizing the risk of environmental contamination to ensure that LAB may successfully hinder the growth of foodborne pathogens. Several foodborne hazards have never been properly controlled via lactic acid fermentation, posing a significant risk to health if they remain in the food. For example, entero hemorrhagic *E. coli* shows acid resistance patterns and can potentially survive specific fermentation processes. Yogurt and fermented meat were identified as potential sources of entero hemorrhagic infections caused by *E. coli*. Food and water-based viruses, which are commonly responsible for causing gastroenteritis, can withstand highly acidic conditions. As an example, it has been demonstrated that Simian rotavirus can withstand high levels of acidity even after being stored within model-fermented foods for 24 hours (Millet et al., 2010).

Considering this, it is crucial to not solely depend on the fermentation process itself to eliminate or decrease risks to safe levels. Ensuring safety is crucial, and the process should be designed to prevent (i) any form of contamination; (ii) minimize or mitigate the risks as much as possible; and (iii) Regulate the proliferation of harmful microorganisms. Due to this reason, it is often necessary to integrate fermentation steps with other procedures operations like soaking, cooking, and similar procedures. Lactic acid fermentation has a minimal impact on anti-nutritional factors, including protease inhibitors and lectins. To eliminate these factors, it is recommended to employ additional process operations like heat treatment (Kunze et al., 2014).

There are also certain considerations regarding the organic acids that generated over lactic acid fermentation. Lactic acid produced by microorganisms is typically a combination of L (+) and D (-) isomers. Humans are unable to metabolize D (-) lactic acid. Excessive production and consumption of D (-) lactic acid can lead to acidosis, which disrupts the balance of acid and alkali in the bloodstream. There is limited knowledge regarding the potential harm of D (-) lactic acid in malnourished or sick children, there is a lack of data on the D (-) lactic acid levels in fermented foods made at home. Additional research is required for better understanding of the function of D (-) lactic acid to create strategies for regulating its production (Sakata, 2019).

There is a potential concern of fermented foods regarding histamine poisoning. While histamine poisoning is often linked to the intake of scombroid-type fish like tuna and mackerel, it is important to mention that some fermented foods, such as wine, soy sauce, and particularly cheese, may also present a potential hazard. Poisoning can occur when there is an excessive build-up of histamine such as tyramine or other biogenic amines, and phenylethylamine. This build-up is a result of the metabolic activity of decarboxylase-positive LAB, like *Lactobacillus biichneri*. However, effectively managing the issue can be achieved by implementing and adhering to strict manufacturing protocols during the production processes. For example, in the production of cheese, careful attention is given to factors like the cleanliness of the milk and the temperature at which it is stored. This is done to reduce the chances of histamine formation and ensure a high-quality product (Giraffa & Carminati, 2008).

8.3. IMPORTANCE OF FOOD FERMENTATION IN PUBLIC HEALTH

Learning Objectives:

- Understand the role of food fermentation in enhancing food safety and nutrition for public health.

Foodborne diseases pose a significant public health challenge globally, affecting nations at various stages of development. Industrialized countries have effectively addressed numerous foodborne infections by using their higher standards of living, access to clean water and sanitation, and advanced food processing and preservation technologies. As a result, there has been a significant decrease in the occurrence of various foodborne infections, including typhoid, fever, cholera, and shigellosis. However, foodborne diseases continue to be a significant public health issue, with statistics suggesting that a considerable portion of the population might get a foodborne illness each year in these countries (Rastogi et al., 2022).

The problem is primarily experienced in the developing world. The lack of available statistics on the incidence of dietary-related illnesses makes it difficult to assess the situation. However, the elevated incidence of diarrheal illnesses in these regions, especially among young children and infants, suggests that there may be an underlying issue with food safety. Approximately 4,000 million incidents of diarrhea were projected to occur globally in 1997. Every year, millions of children within the age of five suffer from episodes of diarrhea, which tragically leads to the loss of countless young lives. Diarrheal diseases have a significant impact on child mortality rates, as it is closely linked to malnutrition, which is a major contributing factor. Every year, approximately 13 million children within the age of five lose their lives due to the devastating impact of malnutrition (Voidarou et al., 2020).

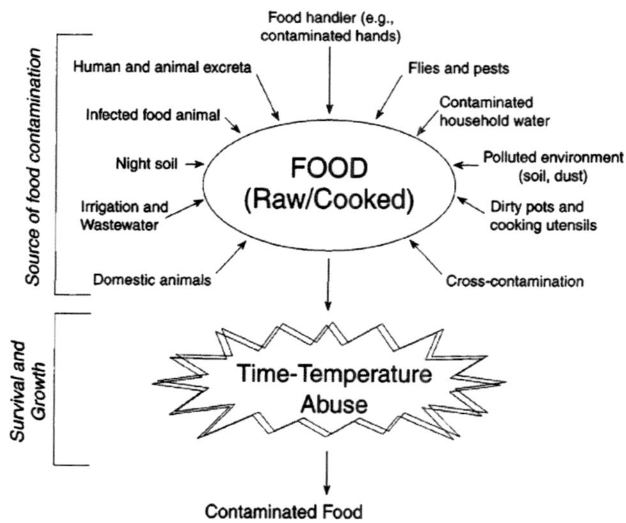

Figure 8.2. *Illustration of sources of food contamination.*

Source: R. Adams, Creative Commons License.

Foodborne diseases may be caused by a wide range of etiological agents, such as bacteria, viruses, and parasites. These agents have been thoroughly examined in separate chapters, providing a comprehensive understanding of their characteristics (Bell et al., 2018).

There are various sources of food contamination, such as contaminated water, waste, airborne particles, insects, animals, unclean utensils, and improper handling by food workers. Contaminants can be found in raw foods due to presence of pathogens or using infected animals in food production. In addition, there is an increased risk of contamination from other ingredients during the preparation process. One of the significant contributors in contamination of food is the improper handling and storage of food, which allows pathogens to thrive and produce harmful toxins (Figure 8.2) (Motarjemi, 2002).

Furthermore, food can serve as a medium for chemical hazardsoccurring naturally or due to inadequate agricultural practices or environmental pollution. This extends beyond just the transmission of agents causing diarrheal diseases. The severity of chemical hazards can vary, ranging from immediate acute intoxication to potential long-term health issues like cancer and various other chronic diseases. Food can sometimes contain substances that hinder the digestion, absorption, or other metabolic processes of nutrients. These substances include enzyme inhibitors, phytates, lectins, and polyphenols (Wilburn & Ryan, 2017).

Food processing technology is utilized for a multitude of purposes. There are various ways in which fermentation can be utilized to improve the quality and safety of food. It can make food easier to digest and more palatable while extending its shelf life. Additionally, fermentation can enhance the nutritional value of food, making it a beneficial process in the food industry. Food fermentation has been found to possess several characteristics, making it a significant food technology. Nevertheless, the utilization of fermentation in enhancing the nutritional value of various food products, especially complementary foods, as well as its ability to preserve foods and inhibit the growth of harmful microorganisms, highlights its significant contribution to public health (Bullerman, 1981) (Figure 8.3).

Fermentation is an essential process that plays a vital role in promoting food safety and minimizing the occurrence of diarrheal diseases, particularly in less developed areas. In developing countries, socioeconomic constraints often hinder the implementation of basic food hygiene practices, which are crucial in preventing diarrheal diseases. These constraints include limited access to safe water, insufficient knowledge and resources for proper preparing and storing food (such as refrigeration and fuel for heating), and time constraints that prevent accurate meal preparation. Due to various constraint, certain households, especially those with limited financial resources, may need help implementing crucial food safety practices.

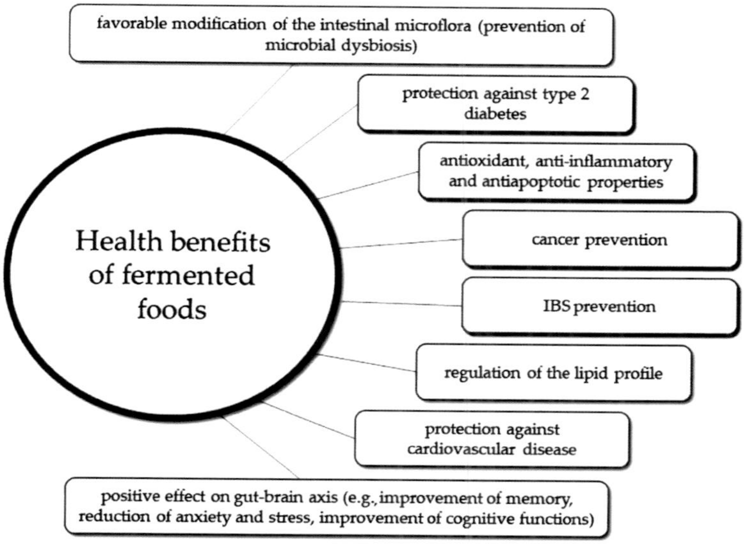

Figure 8.3. *Illustration of health benefits of consuming fermented foods.*

Source: Małgorzata Ziarno, Creative Commons License.

These practices include providing babies with freshly prepared meals, properly storing food at appropriate temperatures, reheating stored food, and more (Durack et al., 2008).

Therefore, fermentation offers a cost-effective method for preserving food and preventing the growth of harmful bacteria, even in situations where refrigerators or other forms of safe storage might not be accessible. Additionally, it has the ability to improve the nutritional value of certain food products (Durack et al., 2008). In certain African nations, this technology is utilized specifically in the production of food for young children and infants. In countries like Nigeria, Kenya, the United Republic of Uganda, and Tanzania is a region in which it is a common practice to feed infants with fermented cereals or root crop products. Additionally, fermentation is employed in the production of various beverages. Once again, in regions where the purity of the water cannot be guaranteed, fermentation processes play a crucial role in minimizing the potential for waterborne diseases (Steinkraus, 1994).

8.4. PRACTICAL INTERVENTION TO ENHANCE SAFETY OF FERMENTED FOODS

Learning Objectives:

- Understand how HACCP principles can be applied to enhance safety in fermented food production.

To fully harness the advantages of fermentation while mitigating any potential risks, it is crucial to thoroughly analyze the fermentation process and devise a comprehensive strategy that addresses and manages hazards associated with each stage of production. Additional instances of implementing the HACCP concept about fermented African food items have been documented in various other publications. Here, one will explore another Indigenous African food called togwa to demonstrate how the theory of HACCP may be applied to assess the safety of a food manufacturing or preparation process. This analysis helps identify potential safety issues and determine effective control measures (Marco et al., 2021).

This study demonstrates that the method of togwa preparation, explained here, results in a product with a high level of risk. There is a potential risk associated with the addition of "power flour" (an ARF derived from germinated seeds) after the cooking process, as it may be contaminated. If fermentation is unsuccessful, it can lead to the growth of unwanted bacteria and the survival of acid-resistant pathogens. Are there any potential methods to mitigate risks, such as expediting fermentation through back slopping or implementing a starter culture? It is crucial to exercise utmost caution to avoid any contamination after cooking. The HACCP examination of this product highlights several critical control points (CCPs) that demonstrate a significant level of concern. (i) There are valid concerns regarding the food's intended audience of infants and young children. These individuals, being more vulnerable to infections, necessitate meticulous attention to the preparation of their meals; (ii) this study has taken into account a variety of hazards; and (iii) The preparation can be done in a home environment with simple conditions. In this specific setting, it is of utmost importance to carefully regulate multiple steps in the process of preparing to avoid or address any potential contamination (Holck et al., 2017).

Another example demonstrates a HACCP study conducted to create a classic cheddar cheese. Cheddar cheese has been linked to cases of salmonellosis and *Staphylococcus* aureus intoxication. In a single outbreak, the issue arose from a failure to properly implement the HACCP system. These issues arose due to inadequate pasteurization or a failure to take appropriate measures when monitoring results indicated a failure to meet crucial thresholds.

Both examples highlight the significance of maintaining proper hygiene practices as a necessary step in producing fermented foods. This is especially important in preventing any form of contamination. When it comes to cheddar cheese, it is evident after the pasteurization process, there are no further measures in place to eliminate pathogens. Thus, maintaining good hygiene practices (GHP) is crucial to avoid any potential recontamination (Nout & Motarjemi, 1997).

8.4.1. HACCP Study of Togwa

8.4.1.1. Product Description

Togwa is a popular beverage among various tribes in Tanzania, with each tribe having its unique methods of production. The process requires combining cooked cereals, such as maize or sorghum, with germinated cereals like finger or bulrush millet or sorghum. The duration of fermentation can vary, typically lasting around 912 hours. However, this timeframe may be influenced by factors such as the use of an old batch as a starter culture.

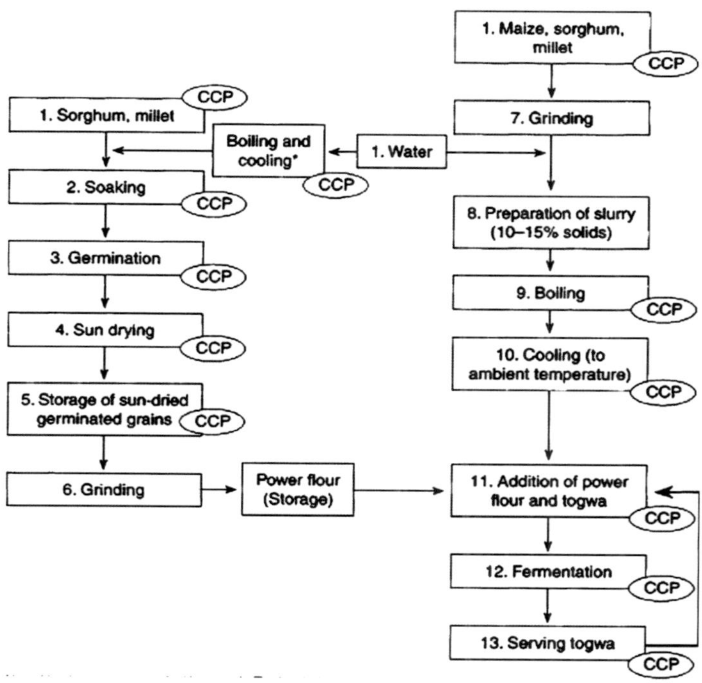

Figure 8.4. *Illustration of the flow diagram of togwa.*

Source: J. Robert Nout, Creative Commons License.

The fermentation process typically takes around 69 hours when using older batches or expired flour. Fermented gruels within Tanzania typically have a short shelf life of one to two days. After a certain period, the product becomes excessively sour or emits an unpleasant odor. Traditionally, fermented gruels have been administered to children under the age of five. Diagram illustrating the flow here is a flow diagram of togwa, as depicted in Figure 8.4.

Risks are taken into account in this particular context encompass various factors such as biological elements (such as bacteria, viruses, and parasites), chemical substances (such as contaminants and mycotoxins), and physical agents (Kussaga et al., 2014).

8.4.1.2. Identification of Hazards, Control Measures, and CCPs

Toxins, such as aflatoxin created by molds, and agrochemicals, pose significant risks in maize, millet, and sorghum. To prevent harmful mold growth during storage, storing the raw material in suitable conditions is important. Limiting the storage time when the ambient temperature and humidity are high is advisableIn case of agrochemicals, there is limited control for households, who must rely on suppliers to provide reassurance about the safety of the products. Preventing unintentional contamination of grains over storage with agrochemicals is crucial.

1. **Raw Material:** It is crucial for households to thoroughly wash the raw material as it could have foreign matter like stones and insect fragments. Unfortunately, the process does not currently include a step to eliminate any pathogens that may have been introduced using raw materials like water, sorghum, or millet. Furthermore, it is important to note that pathogens can potentially multiply over the soaking and germination phases and may even persist after the sun-drying process. Ensuring the safety of the water utilized for power flour preparation is crucial to minimize contamination. The healthy quality of the energy flour is of utmost importance in ensuring the safety of the final product. It is crucial to note that no step in the process guarantees the elimination of acid-resistant pathogens once a power flour is incorporated. As a result of the HACCP research, it was recommended to incorporate a boiling step during the production of power flour. This measure aims to enhance water safety and reduce the risk of power flour contamination (Banwo et al., 2022).
2. **Soaking:** Bacterial proliferation will take place during the process of soaking. Utilizing safe water can reduce the overall bacterial count. Whenever feasible, it is advisable to conduct soaking at moderate temperatures to reduce the proliferation of germs.

3. **Germination:** The process typically occurs in thin layers, which are then placed on mats or leaves. These layers are then covered with additional leaves, mats, or gunnysacks to prevent excessive dehydration. Occasionally, it is necessary to aerate and mix the material while monitoring and adjusting the level of grain moisture. Additional contamination may arise from the presence of mats and the growth of microorganisms during this phase. Microbial growth, especially in molds that produce toxins, can also take place. It is advisable to conduct germination in cool conditions to minimize microbial growth, if feasible.
4. **Sun Drying:** In this stage, the cover is taken away and the sprouted grain is placed out on mats or bamboo trays for drying under the warm sun. It might require a few days. If proper precautions are not taken, various contaminants and foreign substances have the potential to contaminate the sprouted grains. Furthermore, inadequate drying can result in a product that is prone to microbial instability during storage. Ensuring proper drying is essential for maintaining the integrity of the product.
5. **Storage of Sundried Germinated Grains:** Potential risks during this stage include the introduction of contaminants, such as rodents and different animals during storage, as well as the growth of harmful microorganisms, especially toxigenic molds if germinated grains do not get adequately dried or are exposed to high humidity.
6. **Grinding to Power Flour:** During this process, there is a possibility of contaminants being introduced into the product. It is important to prioritize good hygiene practices by utilizing clean and well-maintained equipment whenever possible. However, it is very unlikely that this step will pose any significant health risks.
7. **Preparation of Slurry:** Aside from ensuring the hygiene of water and the pristine condition of utensils, there are no significant risks associated with this particular stage. Since the pathogens that might have been introduced during this step will be killed during the subsequent boiling, it is households need to regard it as a critical control point (CCP). However, households need to prioritize the use of clean utensils and clean water for as long as possible within a GHP.
8. **Boiling:** It is important to ensure thorough boiling to fully gelatinize all starch. It is crucial to ensure the elimination of almost all pathogens, although some bacterial spores may remain.
9. **Cooling:** It is advisable to keep the pot covered to prevent any

contamination from outside sources. Efficient and rapid cooling is crucial in this context. Extended periods of cooling can create favorable conditions for the growth of bacterial spores. When preparing large quantities of togwa, one way to decrease the cooling time is by dividing it into smaller portions.

10. **Incorporation of Fortified Flour and Togwa:** To initiate the fermentation process, two materials can be incorporated: activated flour and pre-fermented togwa. The extent of pollution resulting from power wheat varies and poses difficulties in its management. There is a specific focus on viruses and other agents that can withstand acidic conditions in this context. The togwa culture starter has a high level of acidity, which means it will only have contaminants that can withstand acidic conditions. Through the process of fermentation, acid-sensitive pathogens can be effectively eliminated. Nevertheless, acid-tolerant pathogens have the potential to survive. Producing high-quality flour in a clean environment can help reduce the risk of togwa contamination. Nevertheless, if a thorough reheating step is not implemented, there is a possibility that acid-tolerant pathogens may be present in the final product.

11. **Fermentation:** In the process of fermentation, there is typically a rapid increase in the population of lactic acid bacteria (LAB). This is evident from the relatively quick fermentation process needed to achieve the desired level of acidity. Ensuring a swift fermentation process is crucial in eliminating acid-sensitive pathogens, inhibiting bacterial growth, and preventing the production of toxins. When togwa is added to the fermentation process, it can enhance the overall outcome, as long as it does not introduce any acid-resistant pathogens.

12. **Serving:** It is crucial to maintain strict hygiene practices to prevent the reintroduction of pathogens into togwa through unclean hands and utensils. As a result, it is crucial to thoroughly cleanse them using clean water. The level of contamination in the final product of togwa can vary depending on the hygiene precautions taken during preparation. However, it is important to note that there isno specific control point in place to guarantee the elimination of acid-resistant pathogens. Ensuring proper reheating is crucial for maintaining the quality of the final item. It is important to take the potential impact on texture and other factors, as this could result in a final product that is not well-received by consumers.

SUMMARY

- The chapter explores several key topics related to fermentation. It begins by discussing the benefits of lactic acid fermentation, emphasizing its role in food preservation and enhancement of nutritional value.
- The chapter also addresses the pitfalls associated with fermentation processes, including the risks of contamination and spoilage.
- Importantly, it highlights the significance of food fermentation in public health, underscoring its potential to improve food safety and contribute to dietary diversity. Practical interventions to enhance the safety of fermented foods are examined, focusing on strategies to mitigate health risks and ensure product quality.
- Additionally, the chapter presents a case study using Hazard Analysis Critical Control Point (HACCP) principles in the analysis of Togwa, a traditional fermented beverage.

REVIEW QUESTIONS

1. Explain the benefits of lactic acid fermentation in food processing. Provide examples of foods where lactic acid fermentation is commonly used and how it enhances their properties.
2. What are the main pitfalls associated with fermentation processes? Discuss how these pitfalls can affect food safety and quality.
3. How does food fermentation contribute to public health improvement? Provide specific examples of health benefits associated with consuming fermented foods.
4. Describe practical interventions that can be implemented to enhance the safety of fermented foods. Discuss why each intervention is effective in mitigating risks.
5. Explain the application of Hazard Analysis Critical Control Point (HACCP) principles in the study of Togwa. What specific hazards are addressed, and how are they managed throughout the fermentation process?

MULTIPLE CHOICE QUESTIONS

1. What is the primary benefit of lactic acid fermentation?

a. Increased alcohol content

b. Enhanced food safety

c. Extended shelf life

d. Decreased nutritional value

2. Which of the following is a potential pitfall of fermentation?

a. Increased nutrient content

b. Risk of contamination

c. Enhanced flavor profile

d. Reduced acidity

3. Food fermentation contributes significantly to:

a. Environmental degradation

b. Increased food wastage

c. Public health improvement

d. Water pollution

4. What are practical interventions discussed to enhance the safety of fermented foods?

a. Increasing fermentation time

b. Using less salt

c. Implementing HACCP principles

d. Adding more sugar

5. The HACCP study of Togwa aims to:

a. Increase fermentation temperature

b. Analyze microbial risks

c. Decrease fermentation time

d. Enhance color and texture

Answers to Multiple Questions

1. (c); 2. (b); 3. (c); 4. (c); 5. (b)

REFERENCES

1. Adams, M. R. (2014). Acids and fermentation. In *Food Safety Management* (Vol. 3, pp. 467–479). Academic Press.
2. Bakri, J. M., Maarof, A. G., & Norazmir, M. N. (2017). Confusion determination of critical control point (CCP) via HACCP decision trees. *International Food Research Journal, 24*(2), 852–855.
3. Banwo, K., Ogunremi, O. R., & Adesulu-Dahunsi, A. T. (2022).

Sustainable Use of Food-Grade Microorganisms in Traditional Fermented Food Production. In *Microbial Biotechnology in Food Processing and Health* (Vol. 3, pp. 237–261). Apple Academic Press.

4. Bell, V., Ferrão, J., Pimentel, L., Pintado, M., & Fernandes, T. (2018). One health, fermented foods, and gut microbiota. *Foods, 7*(12), 195.
5. Bullerman, L. B. (1981). Public health significance of molds and mycotoxins in fermented dairy products. *Journal of Dairy Science, 64*(12), 2439–2452.
6. Chelule, P. K., Mokoena, M. P., & Gqaleni, N. (2010). Advantages of traditional lactic acid bacteria fermentation of food in Africa. *Current Research, Technology and Education Topics in Applied Microbiology and Microbial Biotechnology, 2*, 1160–1167.
7. Durack, E., Alonso-Gomez, M., & Wilkinson, M. G. (2008). Salt: A review of its role in food science and public health. *Current Nutrition & Food Science, 4*(4), 290–297.
8. Fadlillah, H. N., Nuraida, L., Sitanggang, A. B., & Palupi, N. S. (2021). Production of antioxidants through lactic acid fermentation: Current developments and outlook. *Annals of the University Dunarea de Jos of Galati Fascicle VI: Food Technology/Analele Universității Dunărea de Jos din Galați, Fascicula VI: Tehnologia Produselor Alimentare, 45*(2), 987–990.
9. Gernaey, K. V., Lantz, A. E., Tufvesson, P., Woodley, J. M., & Sin, G. (2010). Application of mechanistic models to fermentation and biocatalysis for next-generation processes. *Trends in Biotechnology, 28*(7), 346–354.
10. Gilliland, S. E. (1990). Health and nutritional benefits from lactic acid bacteria. *FEMS Microbiology Reviews, 7*(1–2), 175–188.
11. Giraffa, G., & Carminati, D. (2008). Molecular techniques in food fermentation: Principles and applications. *Molecular Techniques in The Microbial Ecology of Fermented Foods, 7*, 1–30.
12. Holck, A., Axelsson, L., McLeod, A., Rode, T. M., & Heir, E. (2017). Health and safety considerations of fermented sausages. *Journal of Food Quality, 2017*(1), 9753894.
13. Ibrahim, A., Yousif, E., ALShukry, A., & Al-Zuhairi, A. (2016). Hazard analysis and critical control point HACCP system. *Iraqi National Journal of Chemistry, 16*(3), 52–60.
14. Isenring, J., Bircher, L., Geirnaert, A., & Lacroix, C. (2023). In vitro human gut microbiota fermentation models: Opportunities, challenges, and pitfalls. *Microbiome Research Reports, 2*(1), 98–100.

15. Kang, K. H. (2013). Health benefits of lactic acid bacteria. *Current Topics in Lactic Acid Bacteria and Probiotics, 1*(1), 1–8.
16. Khubber, S., Marti-Quijal, F. J., Tomasevic, I., Remize, F., & Barba, F. J. (2022). Lactic acid fermentation as a useful strategy to recover antimicrobial and antioxidant compounds from food and by-products. *Current Opinion in Food Science, 43*, 189–198.
17. Kunze, M., Roth, S., Gartz, E., & Büchs, J. (2014). Pitfalls in optical on-line monitoring for high-throughput screening of microbial systems. *Microbial Cell Factories, 13*, 1–21.
18. Kussaga, J. B., Jacxsens, L., Tiisekwa, B. P., & Luning, P. A. (2014). Food safety management systems performance in African food processing companies: A review of deficiencies and possible improvement strategies. *Journal of the Science of Food and Agriculture, 94*(11), 2154–2169.
19. Lee, C. H. (1997). Lactic acid fermented foods and their benefits in Asia. *Food Control, 8*(5–6), 259–269.
20. Levin, D. B., Pitt, L., & Love, M. (2004). Biohydrogen production: Prospects and limitations to practical application. *International Journal of Hydrogen Energy, 29*(2), 173–185.
21. Marco, M. L., Sanders, M. E., Gänzle, M., Arrieta, M. C., Cotter, P. D., De Vuyst, L., & Hutkins, R. (2021). The International Scientific Association for Probiotics and Prebiotics (ISAPP) consensus statement on fermented foods. *Nature Reviews Gastroenterology & Hepatology, 18*(3), 196–208.
22. Mathur, H., Beresford, T. P., & Cotter, P. D. (2020). Health benefits of lactic acid bacteria (LAB) fermentates. *Nutrients, 12*(6), 1679.
23. Mazahreh, A. S., & Ershidat, O. T. M. (2009). The benefits of lactic acid bacteria in yogurt on the gastrointestinal function and health. *Pakistan Journal of Nutrition, 8*(9), 1404–1410.
24. Millet, S., Van Oeckel, M. J., Aluwé, M., Delezie, E., & De Brabander, D. L. (2010). Prediction of in vivo short-chain fatty acid production in hindgut fermenting mammals: Problems and pitfalls. *Critical Reviews in Food Science and Nutrition, 50*(7), 605–619.
25. Motarjemi, Y. (2002). Impact of small-scale fermentation technology on food safety in developing countries. *International Journal of Food Microbiology, 75*(3), 213–229.
26. Nout, M. J. R., & Motarjemi, Y. (1997). Assessment of fermentation as a household technology for improving food safety: A joint FAO/WHO workshop. *Food Control, 8*(5–6), 221–226.

27. Obadina, A. O., Oyewole, O. B., Sanni, L. O., Tomlins, K. I., & Westby, A. (2008). Identification of hazards and critical control points (CCP) for cassava fufu processing in South-West Nigeria. *Food Control, 19*(1), 22–26.

28. Rastogi, Y. R., Thakur, R., Thakur, P., Mittal, A., Chakrabarti, S., Siwal, S. S., & Saini, A. K. (2022). Food fermentation–Significance to public health and sustainability challenges of modern diet and food systems. *International Journal of Food Microbiology, 371*, 109666.

29. Rodríguez, L. G. R., Gasga, V. M. Z., Pescuma, M., Van Nieuwenhove, C., Mozzi, F., & Burgos, J. A. S. (2021). Fruits and fruit by-products as sources of bioactive compounds. Benefits and trends of lactic acid fermentation in the development of novel fruit-based functional beverages. *Food Research International, 140*, 109854.

30. Sakata, T. (2019). Pitfalls in short-chain fatty acid research: A methodological review. *Animal Science Journal, 90*(1), 3–13.

31. Sharma, R., Garg, P., Kumar, P., Bhatia, S. K., & Kulshrestha, S. (2020). Microbial fermentation and its role in quality improvement of fermented foods. *Fermentation, 6*(4), 106.

32. Sionek, B., Szydłowska, A., Küçükgöz, K., & Kołożyn-Krajewska, D. (2023). Traditional and New Microorganisms in Lactic Acid Fermentation of Food. *Fermentation, 9*(12), 1019.

33. Steinkraus, K. H. (1994). Nutritional significance of fermented foods. *Food Research International, 27*(3), 259–267.

34. van de Sande, T. (1997). Socio-economic pitfalls of enhancing indigenous capabilities in household fermentation. *Food Control, 8*(5–6), 303–310.

35. Verbelen, P. J., De Schutter, D. P., Delvaux, F., Verstrepen, K. J., & Delvaux, F. R. (2006). Immobilized yeast cell systems for continuous fermentation applications. *Biotechnology Letters, 28*, 1515–1525.

36. Voidarou, C., Antoniadou, M., Rozos, G., Tzora, A., Skoufos, I., Varzakas, T., & Bezirtzoglou, E. (2020). Fermentative foods: Microbiology, biochemistry, potential human health benefits and public health issues. *Foods, 10*(1), 69.

37. Waché, Y., Do, T. L., Do, T. B. H., Do, T. Y., Haure, M., Ho, P. H., & Chu-Ky, S. (2018). Prospects for food fermentation in South-East Asia, topics from the tropical fermentation and biotechnology network at the end of the AsiFood Erasmus+ Project. *Frontiers in Microbiology, 9*, 2278.

38. Wilburn, J. R., & Ryan, E. P. (2017). Fermented foods in health promotion and disease prevention: An overview. *Fermented foods in health and disease prevention, 5*(7), 3–19.

39. Ye, H., Wang, J., Shi, J., Du, J., Zhou, Y., Huang, M., & Sun, B. (2021). Automatic and intelligent technologies of solid-state fermentation process of Baijiu production: Applications, challenges, and prospects. *Foods, 10*(3), 680.

INDEX

E

F

FSC
www.fsc.org
MIX
Papier | Fördert gute Waldnutzung
FSC® C083411

Kolibri 360 GmbH
Ferdinand-Jühlke-Straße 7
99095 Erfurt, Deutschland
produktsicherheit@kolibri360.de